Experimentation in Biology

TERTIARY LEVEL BIOLOGY

A series covering selected areas of biology at advanced undergraduate level. While designed specifically for course options at this level within Universities and Polytechnics, the series will be of great value to specialists and research workers in other fields who require a knowledge of the essentials of a subject.

Other titles in the series:

Methods in Experimental Biology	Ralph
Visceral Muscle	Huddart and Hunt
Biological Membranes	Harrison and Lunt
Comparative Immunobiology	Manning and Turner
Water and Plants	Meidner and Sheriff

TERTIARY LEVEL BIOLOGY

Experimentation in Biology

An Introduction to Design and Analysis

W. J. RIDGMAN, M.A. (Cantab.)

Lecturer in Field Experimentation
Department of Applied Biology
University of Cambridge

A HALSTED PRESS BOOK

John Wiley and Sons
New York

Blackie & Son Limited
Bishopbriggs
Glasgow G64 2NZ

450 Edgware Road
London W2 1EG

Published in the U.S.A. by
Halsted Press,
a Division of John Wiley and Sons Inc.,
New York

© 1975 W. J. Ridgman
First published 1975

All rights reserved.
No part of this publication may be reproduced,
stored in a retrieval system, or transmitted,
in any form or by any means,
electronic, mechanical, recording or otherwise,
without prior permission of the Publishers

Library of Congress Cataloging in Publication Data
Ridgman, W   J
Experimentation in biology.

(Tertiary level biology)
Bibliography: p.
Includes index.
1. Biometry. I. Title.
QH323.5.R53   574.01'8   75-38516
ISBN 0-470-15216-8

Printed in Great Britain by
Thomson Litho Ltd., East Kilbride, Scotland

Preface

THE NEED FOR SOME STUDY OF BIOMETRY IN COURSES OF BIOLOGY IS NOW widely accepted. It is unfortunate that many students of biology do not take a delight in mathematics for its own sake. Thus anything remotely mathematical, such as "statistics", tends to be studied rather grudgingly as a separate entity, and to be used equally grudgingly to placate editors of journals who ask for evidence to support a young biologist's bright idea.

There are many excellent books on statistics, but most of them start from the view that the reader wishes to learn statistics. Cambridge was one of the first universities to introduce lectures on statistics orientated specifically towards biologists. Udney Yule was appointed lecturer in 1912, and he collaborated with F. L. Engledow, now Sir Frank Engledow, Emeritus Drapers Professor of Agriculture, to produce one of the first books combining theory with practice: *Principles and Practice of Yield Trials*, published by the Empire Cotton Growing Corporation. From there on, teaching in the Faculty of Agriculture and now in the Department of Applied Biology has always been carried out as a combined operation by two lecturers: one a biologist who approaches his work from the need to establish biological truths, and the other a mathematician who can supply the rigour necessary to ensure that mathematical devices are used correctly. The present book has grown out of this arrangement and is the biologist's contribution. It is not meant to be a recipe book to which one turns to "find a good design"; the design of any experiment depends on the questions being asked and the material available, and must therefore be worked out carefully for each occasion. Rather it is hoped that it forms a useful study in itself for biologists who wish to make experiments, and for those who wish to appreciate the experiments of others. They should then turn to more mathematically orientated books to obtain the rigour if they wish to become biometricians.

The problems covered, and many of the examples, have been brought to me by past and present colleagues in the Department. From them and from some 1500 students who have attended my courses I have received unstinting help in pin-pointing the areas in which biologists find difficulty. For this I am most grateful.

W. J. RIDGMAN
Department of Applied Biology
University of Cambridge

Contents

		Page
Chapter 1.	INTRODUCTION	1

Modes of thought in biological study. Observation, hypothesis, experiment. Testing, estimating. Belief, ideas of certainty. Variation. Quality of biological data, nominal, ordinal, interval scale; definitions, conversion, relationship in terms of information obtained; some examples. Derived values, ratios, percentages, their meaning and possible ambiguities.

Chapter 2. THE EXPERIMENTAL METHOD — 8

Defining a population. Sampling. Unbiased samples, systematic samples, random samples. The process of randomization. Designing a simple experiment; null hypothesis, assigning the objects to the treatments. Avoiding confusion of effects while making the experiment.

Chapter 3. STATISTICAL ANALYSIS — 17

Data from the experiment designed in Chapter 2. Condensing an array of values. How a reasonable man might picture a population. The statistical model and its relation to the sample and the population. Calculations required for sum of squares, degrees of freedom, mean square, standard errors and coefficient of variation.

Chapter 4. TESTING AND SETTING CONFIDENCE LIMITS — 35

Students' t-test. Least significant difference. Confidence limits; their relation to testing. Consideration of the factors involved in the t-test. Single factor design, the assumptions made. Fisher-Behrens test; Mann-Whitney's U.

Chapter 5. EXPERIMENTS WITH MORE THAN TWO TREATMENTS   54

Testing if all treatments represent the same population, analysis of variance, F-test, Kruskal-Wallis procedure. Tests within a set of treatments; *a posteriori* testing. Duncan's multiple range test. Individual contrasts nominated before the experiment starts, comparisons with a control, Dunnett's t-test, the number of replications for the control. Orthogonal contrasts, three treatments, four treatments.

Chapter 6. FACTORIAL ARRANGEMENTS OF TREATMENTS   81

Two factors, each at two levels. Interactions. A use of χ^2. Advantages of factorial arrangements. Design and analysis of a 2^3 arrangement, interpretation and display of the results.

Chapter 7. USE OF ORTHOGONAL POLYNOMIALS, INTERACTIONS AND REGRESSION   101

Derivation of orthogonal polynomials, testing, drawing smooth curves. Cases with more than one treatment factor. Some other sorts of interaction. Regression, estimating the parameters. Correlation coefficient. Kendall's rank correlation.

Chapter 8. ERROR VARIATION AND ITS CONTROL   130

The single-factor design. Hierarchical designs, confidence limits for variances, two-factor designs, random and fixed-effect factors, randomized blocks. Alternative non-parametric procedures, Wilcoxon's signed-rank test, Sign-test, Friedmann's procedure. Three-factor designs. Latin squares.

Chapter 9. CONFOUNDING   155

Split-plot design. Confounding in 2^n designs, partial confounding. Use of high-order interactions as error, single replicates in incomplete blocks. Fractional replication of 2^n designs. Confounding in 3^n designs. Examples of confounding where factors have more than three levels and when the number of levels is not the same for all factors.

Chapter 10. LATTICE DESIGNS AND NON-ORTHOGONAL ANALYSIS   195

An example of a lattice square. Treatments applied in sequence in a Latin square. Biological problems of non-orthogonal analysis.

Chapter 11. ANALYSIS OF COVARIANCE 214
> Controlling error variation by taking measurements before the experimental treatments are applied.

FURTHER READING 224
APPENDICES 225
> Random digits. Values of t and F. Values of χ^2. Values for Duncan's multiple-range test. Values of t (Dunnett).

INDEX 231

CHAPTER ONE

INTRODUCTION

Serious study of almost any subject usually involves three aspects: collecting evidence, processing the evidence, and drawing a conclusion. This is certainly true in biology, and it is worth considering the tools which are required to carry out these procedures before embarking on such a study.

Collecting the evidence includes reading the published work of others, careful observation all day and every day (which of course is no hardship to a biologist, because almost everything he sees is concerned with his subject) and making experiments, i.e. deliberate appeals to experience. Some will argue that the last of these is not essential, as the necessary experience can be gained by observing diligently the organism being studied in its natural conditions. This may be true of certain studies when considered in the long run, but it is a slow process, and there are situations in which it is not true. For example, if an entomologist has the idea that a certain uncommon chemical will kill aphids, he is very unlikely to see an aphid which has eaten that substance under natural conditions, and will never know if he is right unless he makes an experiment. There are many such situations in biology, where progress beyond present knowledge can be made only by introducing circumstances or conditions which have not been known to exist before, and that means making an experiment.

Processing the evidence includes a wide range of activities, and much of this book is devoted to the methods available. Basically there are two forms of attack. Firstly we can use a deductive approach in which certain observations are taken as true or axiomatic, and then, often with the aid of a mathematical model, we can argue various outcomes quite rigorously. For example, in taxonomy we might use the following simple mathematical logic: if all the As are Bs and all the Bs are Cs, then all the As are Cs; but in general the problem when using this approach in biology is to establish any observations as axiomatic. Living organisms are always changing, and if the basic assumptions have never been challenged, this does not mean that they will not be challenged at some time in the future. Secondly we can adopt what is sometimes called the *empirical approach* of setting up a hypothesis, and then using the available evidence to try to dispute it. Until the hypothesis is disproved, it remains a tenable theory, but of course this method is by no means foolproof, because it requires that all possible challenges are made before the theory can be fully established.

The hypotheses most amenable to this form of attack are those which can be tested in the form of a question answerable by either yes or no. For example, if the hypothesis is: feeding vitamin C to rats will affect their growth rate, we can ask the question: does feeding vitamin C to rats affect their growth rate; and the answer can be only yes it does, or no it does not. Experiments are often used to *test* this kind of hypothesis, though it will be seen later that biometrical methods are such that it is better to put the hypothesis in the negative (called a *null hypothesis*) which in this case would state: feeding vitamin C to rats does not affect their growth rate. However, the question derived from this is really the same.

Experiment is also useful for answering questions in the form: how much. For example: how much faster do rats grow when fed vitamin C. Biometrical methods enable this *effect* to be estimated. In fact the two situations require the same sort of arithmetic, because to carry out any *testing* it is necessary to *estimate* a quantity representing a difference and to see if this quantity differs from zero. It is, however, important to distinguish between them when designing an experiment. The setting up of an experiment to estimate something without any prior hypothesis, then looking at the data, deriving a hypothesis from them and using the same data to test the hypothesis, is a common form of self-deception. Of course, we may make use of data from experiments to derive hypotheses, but these should then be tested quite separately in new experiments set up for that purpose. It should be stressed that the form of analysis of any experiment concerned with testing should be dictated by the hypotheses being tested, not by the appearance of the data.

The third operation of any study is the drawing of conclusions. In studies which adopt the deductive approach, the conclusions can be quite certain, provided the assumptions or axioms are true. In geometry, for instance, it was customary to write Q.E.D. (*quod erat demonstrandum*) at the end of a theorem, but it should be remembered that the theorem is proved only within the bounds set by the assumptions, though these are seldom debated. In biology, the assumptions or axioms are usually much less acceptable, and it is frequently necessary to state them when drawing conclusions. Further, biologists have to draw conclusions about organisms which at present do not exist, and so it is not surprising that in everyday life people are loath to make categorical statements about anything biological. Even the statement "as surely as night follows day" leaves a loophole, but more common are expressions such as "nine times out of ten this happens; therefore I believe that it is going to happen on this occasion". The latter statement is the form of statistics used by many people (even by those scientists who scorn statistics) and, put into a more rigorous form, it leads to a powerful weapon for the study of biology. It must, however, be realized that the likelihood to which we are referring is the likelihood that we are right in drawing the conclusion. It is our belief which is in question, and so, in addition to obtaining experimental evidence on which to draw conclusions, it is necessary to present evidence for the strength of our belief. Statistical procedures allow this to be done.

A difficulty in biological study is that organisms are patently very variable; in fact every living thing is unique, and no two members of any species are exactly alike. This is also true of non-living objects but, in that case, the methods of construction can often be refined to remove much of the variation, until it becomes so small that it does not matter in any argument or any use of the object. However, in living organisms the variation is there as part of the process, hence it is usually referred to as *natural variation*, and any methods of study or experimentation must take account of it. It is for this reason that the study of a single organism, no matter how careful or detailed, can seldom lead to any firm conclusions concerning any other organisms, even if they are of the same species and living under what could be described as the same conditions.

Data from experiments

Natural variation also has a bearing on the quality of the data which can be usefully collected. If mathematical models are to be used in assessing evidence, then some form of numerical data is often desirable. Three forms are common, and their quality should be considered before embarking on any study. The simplest is perhaps that known as *nominal data*, e.g. when assessing the effect of a fungus on grapes we might restrict our data to the presence or absence of a brown spot on the grape and, if we wanted to do arithmetic with the data, presence of a brown spot could be given the value 1 and absence the value 0. We would not claim these to be high-quality data because, although 0 stands for something definite (namely, a grape we can eat) which we could communicate to others without any chance of misinterpretation, 1 could represent very variable conditions: the grape might have a very small spot on it (which few would worry about and would be prepared to buy and eat) or it might be a completely brown and squashy grape which would have to be thrown away. However, such data are all that we can obtain in many instances in biology. An improvement is *ordinal data*, i.e. data in which objects are ranked according to the extent to which they show a certain character. In the case of the grapes, we might want to compare three, and could rank them according to the extent of browning, being able to say without doubt that one had most browning, one had least, and one was intermediate. To introduce arithmetic, we could give them values of 3, 1 and 2 respectively, but it must be remembered that this process does not imply the usual arithmetical rule that $3 - 2 = 2 - 1$, because there is no reason to believe that the grape with most browning had twice as much as the intermediate one. So there are limitations to data of this type. In addition, it is difficult to convey any absolute statement to anyone else; a grape with rank 3 in one set of observations does not necessarily have the same amount of browning as the one ranked 3 in another set. Nevertheless this process, often known as *scoring*, is sometimes the best available.

The best data are measurements on an interval scale. Here we can carry out all the usual arithmetical processes and, when communicating our results to others, we can give them a much more precise description. In the case of the grapes, if we could measure the area of browning, we should have a measure which should distinguish each grape from every other grape, provided we could measure the area to a sufficient degree of accuracy. In fact, in true interval scales it is assumed that no two values are the same. This would be so if they were taken to sufficient decimal places but, in real life, measurements are made only to a certain number of decimal places prescribed for the task. Therefore care is required to measure sufficiently accurately if the assumptions of interval scale are to be used. Criteria for deciding how accurately to measure will be discussed later, but it should be pointed out here that while it is easy to downgrade the quality of data after they have been collected, it is not possible to upgrade it. Thus it is obviously possible to round off and to achieve measures with fewer decimal places. It is also possible to convert measurements on an interval scale to measurements on an ordinal scale. For example, on an ordinal scale the values

could be ranked   1·23   2·73   4·86
                              1   2   3

INTRODUCTION

Notice that $4{\cdot}86 - 2{\cdot}73 = 2{\cdot}13$ is given the same weight as $2{\cdot}73 - 1{\cdot}23 = 1{\cdot}50$ on the ordinal scale: information has therefore been lost in the process. Yet more information would be lost in conversion to a nominal scale where, for example, we take values $\geqslant 2{\cdot}00$ as acceptable (A) and $< 2{\cdot}00$ as unacceptable (U) and get

$$
\begin{array}{ccc}
1{\cdot}23 & 2{\cdot}73 & 4{\cdot}86 \\
U & A & A
\end{array}
$$

for no longer can we distinguish between 2·73 and 4·86.

Nevertheless data which are inherently nominal or ordinal can be collected in such a way that they can be used as interval-scale data in analysis. For example, plants of certain *Phaseolus spp.* usually produce one inflorescence stalk at each node, but certain hormonal treatments will cause them to produce two. Counting inflorescence stalks on a node thus produces nominal data in which only two values, 1 or 2, are possible; but if we take 10 nodes as our unit for counting, the possible values are all integers from 10 to 20, which we could certainly treat as an ordinal scale. Counting 100 nodes, with possible values from 100 to 200, we would be as near a true interval scale as many direct measurements made to three significant figures.

Likewise, combining independent estimates made on an ordinal scale will often produce data which can be treated as if they were measured on an interval scale. This, however, should not be done without careful investigation, particularly if it is desired to draw the conclusion in terms of a particular known scale. An example of the sort of investigation necessary is to be found in Bald (1943),* where the object was to obtain a scale for leaf area of potato plants by scoring. Most experimenters can distinguish 10 categories of size and give them ranks 1 to 10. Bald, however, discovered that the interval between the ranks was not directly related to the area of leaf but to the logarithm of the area, because an observer required a greater absolute difference to distinguish two large leaves than to distinguish two smaller leaves. Thus he could produce a straight-line relationship between score and logarithm of area, and obtain the actual area by taking antilogarithms.

There are many opportunities in biology to use simple measurements and then to transform them into values more easily manipulated or more easily understood. Such occasions include situations where we require a measure such as the weight of a plant during growth, which can be determined directly only by destroying the plant. Then a scoring technique

---

* Reference to page 224 will give further details.

such as that used by Greenwood, Cleaver and Niendorf (1974) is very useful. However, such devices are possible only if there is a one-to-one relationship between the data recorded and the measures sought.

Transformation of data

Similar transformations or combinations are often made with data which are on an interval scale when collected. Here again, interpretation needs care. A common procedure is to express measurements as percentages of something. This often produces one value instead of two but, to illustrate the dangers, consider these two ways of presenting the results of an experiment devised to assess the effects of salinity on germination of two varieties of groundnuts, with a view to recommending which should be chosen for saline soils.

Table 1.1(a)—Germination of groundnuts on saline and non-saline soils (%)

	Non-saline soil	Soil + 10 mg NaCl/kg
Variety A	98	30
Variety B	63	23

(b)—Germination of groundnuts on saline soil, expressed as a percentage of germination on non-saline soil

Variety A	30·6
Variety B	36·5

The claim from Table 1.1(b) would be that variety B was more salt-tolerant than variety A. If, however, variety B consistently germinates only 63 per cent under non-saline conditions, as is shown in Table 1.1(a), this is misleading information for the grower who wishes to get the best stand from a given amount of seed. One of the great dangers of combining data is that some salient facts may be hidden and forgotten by the experimenter himself. This may also affect any extrapolation that a reader might be tempted to make. It is common practice in some commercial areas to make experiments determining how much extra produce may be obtained by a change of practice, such as application of a chemical. Suppose experiments show that application of a chemical increases a yield of barley by 0·3 t/ha, and that the average yield of barley, without chemical, is 3·0 t/ha. It would be true to claim that the chemical had increased yield by 10 per cent but, if that is the only value published, it may lead a grower normally achieving 4·0 t/ha to believe that the chemical increases yield by 0·4 t/ha—a substantial miscalculation (0·1 t being worth £6 at the present time).

In general, it is wise to decide on units or measures before starting an experiment: otherwise it is possible to obtain different answers according to arithmetical manipulations. It would then be possible to change the hypothesis into that which suited the data best. A simple example is the measurement of both food intake and weight gain of animals. If two animals eat a_1 and a_2 kg of food, respectively, resulting in weight increases of b_1 and b_2 kg, respectively, then although

$$\frac{1}{2}\left(\frac{a_1}{b_1} + \frac{a_2}{b_2}\right) \quad \text{would not equal} \quad \frac{a_1+a_2}{b_1+b_2}$$

except under rather special conditions, both might be considered to be estimates of average conversion efficiency, i.e. the amount of food eaten per kilogram live-weight increase. The former would seem preferable in considerations of what an animal does with its food, but the latter is preferable where the aim is to determine how much food to supply to a similar bunch of animals. There are many such problems in biology, such as whether to consider the reciprocal of the measurement, which sometimes makes sense. For example, when various numbers of animals are grazing a given area of grass, it is the number of animals which is recorded, but the important character is how much grass each animal can obtain. That is measured by $\frac{\text{area}}{\text{number of animals}}$ hence the reciprocal of number of animals would be the more useful measure. Again, many processes in biology function in a multiplicative manner, and transforming to logarithms in these cases will make for more sensible calculations. In fact, for ease of interpretation, it would be true to say that the aim is to convert measures into a form which is linearly related to the way in which the effect is understood, so that the effect is measured on a straightforward linear scale. In all cases, the scale must be decided for biological reasons before the experiment is made.

CHAPTER TWO

THE EXPERIMENTAL METHOD

Biologists will readily appreciate that experiments can be made with only a very limited amount of material or only a few organisms. Yet usually the whole object of an investigation is to determine characters which apply to a very large amount of material or to all the members of a particular species of organism. It is a sobering thought that no one is interested in a research worker's experiment except in so far as he himself would obtain the same result if he followed the research worker's recommendations. To achieve this objective, experiments are made on a small amount of material which can reasonably be said to represent the mass of material to which the results are to apply. Before starting any experiment, then, the mass of material or *population*, as it is called in statistics, to which the results are to apply, must be carefully defined. Failure to do so is a common cause of confusion among results of different workers. It is all very well to paint a broad canvas and say, for example, that we are investigating the sensory mechanisms of rodents but, when practical considerations such as time and finance restrict the experiment to a comparison of the effects of two pitch frequencies on the movement of a single strain of laboratory rat, the application of the results is obviously much more restricted. Thus definition of the population is often a two-stage process. We think firstly of the ideal, which is suggested by the problem itself; but we then have to consider practical possibilities and trim the population accordingly. Animal work provides a good example. Often we would like to make remarks about a species living in any habitat at any time; but in many cases practical considerations restrict experimentation to particular strains or breeds living in particular habitats which are accessible at the time when the facilities are available. When reporting the results, it is most important to state clearly the limits to the population which has been investigated.

Having defined the population, it is necessary to ensure that a properly representative portion of it is used in the experiment. The portion used would be known statistically as a *sample* of the population, and the degree of reliance that can be placed on the estimates of the true population values derived from a sample is very much affected by the method of drawing the sample. Practical procedures for sampling are fully discussed in Yates (1960), and an elementary account of the theory can be read in Sampford (1962), but when selecting material for biological experiments one can often only approximate to the ideal. If this is done, care is required when drawing conclusions. In fact the biologist is often faced with the problem that he has started with the only sample available, and must then decide what population this represents, rather than choosing a sample rigorously from an already existing and defined population. Such situations are unavoidable, but the experimenter should be fully aware of the material with which he is experimenting and should make abundantly clear to the readers of his results to what other material the results should apply.

If the whole population can be delineated, various methods of obtaining representative (or *unbiased*) samples can be used. We should consider two. Firstly a systematic sample, which is often used in ecology when the frequency of occurrence of all the species falling on a line drawn across an area might be considered to be representative of the frequency of species in the whole area. Similarly, in a blood count, certain specified squares might be counted in every sample of blood examined. This will be an unbiased sample, provided there is no systematic distribution within the population which is mirrored by the systematic sample itself. For example, suppose an ecologist is sampling old ridge and furrow land where the ridges are 11 yards apart. If he chooses transects at 11-yard intervals parallel to the ridge, he will be sampling only one particular elevation, and so no estimate would be representative of the area as a whole. Here a systematic sample requires to be worked out to ensure that all elevations appear equally, if the result is to be of any use.

In many experiments, the main objection to the use of systematic sampling is that it does not provide the sort of estimate of natural variation which is required, if we are to use the best statistical techniques.

Randomization

The other sort of sample, which is of the utmost importance in experimentation, is a *random sample*. "Randomization" does not mean some haphazard process, but rather a rigorous process of selection in which every member has an equal chance of being selected. Whether a sample is a random one or not is entirely dependent on the method by which it is drawn. When dealing with populations in which members can be enumerated in some way, there are various ways in which a really random sample can be achieved. One which no doubt springs to mind is the use of an electronic randomizer such as ERNIE which selects the winners of Premium Bonds. Such apparatus is most convenient when there are many members from which to draw, but few biologists need such sophisticated apparatus. A more common method used in everyday life is the drawing of numbers from a hat; this is perfectly all right provided that the cards carrying the numbers are all of the same size, are folded in the same way and cannot become stuck together, and provided that there are no places in the hat where cards can lie hidden during successive draws. However, most experimenters use a table of random digits such as is provided in Table I of the Appendix. Random digits have been prepared in various ways and then tested for randomness, and the digits in this table can be relied upon as being random in all directions and in all consecutive groupings. Thus from the table we can read off

single digits across the page: 1, 5, 0, 6, 6, 6, 5, 2, 8, 5, 0, 8;   these are random;

single digits down the page: 1, 4, 9, 2, 5, 8, 3, 2;   these are random;

double digits across the page: 15, 6, 66, 52, 85, 8 (or down);   these are random.

Groups of three or more digits may be used similarly if large numbers are required. Furthermore, they are random no matter where we start or in which direction we go.

Let us use them for a simple example. Suppose we wished to select a random sample of four undergraduates from a class of 10, to carry a heavy calculating machine to the laboratory. Suppose we took a fresh sample each day for a number of days. We should first number the members from 1 to 10 quite arbitrarily. Then to select four the first day, we could start at the top of column 1 of the table and, reading down the column, choose numbers 1, 4, 9 and 2. On the second day we would have the next four numbers, 5, 8, 3 and 2. On the next day, the random digits would be 0, 2, 5, 0. Obviously we take 0 to mean "man number 10", but this member has appeared twice and we have a situation requiring four men; it would not be possible for the total to be made up by the appearance of one man twice. We should therefore discard the second 0 from the sample and take the next digit 6. Then 10, 2, 5 and 6 is the sample for that day. This system is known as *sampling without replacement* (derived from the idea that when drawing numbers from a hat, the cards are not replaced once they have been drawn, and so cannot be drawn twice). It is the type required in most experiments where we require a definite number of objects in the sample. *Sampling with replacement* would be more appropriate to the case of Premium Bonds, where there is no ruling that one bond must not win more than one prize at one draw. If, then, our numbers 10, 2, 5 and 10 were drawn, number 10 would get two prizes.

This method of randomization is applicable no matter how many members there are from which to select. Using a single digit, we can select from any number up to 10, e.g. to select from 7, proceed as before but ignore 8, 9 and 0. If the number of members is greater than 10 but less than 100, use pairs of digits, and so on. Yet another feature of these random digits is that if they are divided by certain constants the remainders are random digits as well. For instance, if we wanted to select our 4 men out of 20, we should have to use pairs of digits, because members like 11 or 19 must be able to appear. If we do this in the straightforward fashion, it will obviously take a long time because, as we go down the column, only one fifth of the numbers we see have anything to do with our problem; numbers from 21 to 100 (00) are of no use to us. If, on the other hand, we divide every number by 20, the remainder is always a number between 1 and 20, provided we take numbers with no remainder (i.e. 20, 40, 60, 80 and 00) as having a remainder of 20. Now since we have the numbers 1 to 100 occurring at random in the table, each of the numbers 1 to 20 has an equal opportunity of selection: 1 can come from 01, 21, 41, 61 and 81, and 18 from 18, 38, 58, 78 and 98, and so on. In generalizing the method, the criterion of equal likelihood of selection for all members must always be considered. Suppose, for example, we wanted to select out of 16, then using the remainder method we could get number 1 from the random digits 01, 17, 33, 49, 65, 81 and 97, i.e. in seven different ways. On the other hand, number 16 could be selected only from 16, 32, 48, 64, 80 and 96, i.e. in six different ways. Our selection would therefore be biased in favour of number 1 compared with number 16, number 1 having a greater chance of occurring. In fact, numbers 1, 2, 3 and 4 can all appear in seven ways, while 5, 6,..., 16 can appear in only six ways, and so we should be far from an unbiased selection. The solution to the problem is simple: we ignore the random digits 97, 98, 99 and 00, and then all members can appear in only six different ways; since the discarded numbers themselves appear at random, we now have an unbiased sample.

The rule then is as follows: to use the remainder method of randomization, first find the largest number in the set which is exactly divisible by the number of objects from which a selection is to be made; discard all numbers greater than this and then write down the remainder formed by dividing the remaining random numbers by the number of objects, counting a remainder of 0 as the number of objects itself.

The idea of randomization, like all statistical concepts, is concerned with lack of bias in the long run, and it must not be expected that each member will appear the same number of times in only a few samples. In the present example, undergraduate number 2 would be carrying the machine on each of the three days, while number 7 would not carry it at all; but if we continued to draw samples indefinitely all undergraduates would carry it the same number of times. This emphasizes the need to ensure that every randomization is independent of any other randomization. In these days when extensive tables of random digits, such as are to be found in Fisher and Yates (1963), and computers with randomizing facilities are available, there is no excuse for using the same selection more than once. Using tables, a biologist would be advised to start at the beginning of his tables when he starts his experimental life, mark off the last digit used on the first occasion, then start at the following digit on the next occasion, and so on, returning to the beginning again when he has used the complete table. Nor should anyone need to use less precise methods, like chasing pigs round a pen, opening the gate and considering the first four out to be a random sample; they may well be a special population of pigs with a desire to get out, and may have many other characters connected with that desire.

Although perfect randomization can be carried out when all the members of the population can be enumerated, there are many biological situations where samples are required to represent large populations, all of which could not be found and certainly could not be used by the experimenter. It is here that approximations have to be made, but it is well to remember that we can make remarks of known reliability about a population of objects only if we get the information from all the objects or from an unbiased sample of them. Using any other sort of sample, as often we must, involves some guesswork, and so we must not be surprised if sometimes the guesses prove to be false.

Designing an experiment

Suppose we wish to know if spraying the plants of a new variety of wheat with a fungicide will affect its yield. Let us assume that at this stage we are not concerned with whether spraying would be profitable in the field, but are content to know if there is any effect when the plants are grown in pots filled with a certain John Innes compost, and subjected to certain environmental conditions in the presence of some known incidence of the fungus. To be able to use statistical methods to test for an effect, we must first make a hypothesis. In all science, a hypothesis is something we think up and believe until some evidence comes along to dispute it. Normally it is not possible to prove a hypothesis completely, no matter how many times it is tested: on the other hand, it is disproved if it fails once. So for experimental purposes hypotheses are made in the negative. They are called *null hypotheses*, and the experiment is one opportunity to dispute the null hypothesis. In the present case, we would use the null hypothesis that "there is no difference in yield between spraying the plants of this variety with fungicide and not spraying the plants". If, after making the experiment, we find this null hypothesis to be no longer acceptable and refuse to believe that there is no difference, then we must believe that there is a difference. Thus the first requirement when designing an experiment is a clear statement of the objectives and, if tests are to be made, the null hypotheses must be stated clearly.

To test the present null hypothesis we must compare two populations:

(1) the population comprising the variety grown according to the stated method and sprayed with fungicide; and
(2) the population comprising the same variety grown in the same way but not sprayed with fungicide.

Some find difficulty in this concept, because they know full well that the first population exists only in the experimenter's imagination at the start of the experiment. However, if the experiment disputes the null hypothesis, it is likely that this population will soon exist and, if the sample has been properly defined, there should be no real difficulty in deciding to what other material the results will apply.

It is seldom possible to use the whole of an existing population in any experiment, and so samples have to be used. Quite intuitively, we would expect the reliability of the results to depend upon the size of sample used, and indeed it will be shown later that we can calculate how many members are required to achieve a given level of confidence in the results obtained. The number of units representing each population is called the number of *replications* in experiments, a complete set of one unit for each of the populations under study being called a *replicate*. Replication serves two functions: firstly it increases the precision of the estimate and, secondly, it is necessary in many experiments for estimation of the natural variation of the material which is used in statistical tests.

In the experiment under discussion, we should first ensure that the seeds used were representative of the variety stated. If it is a new variety bred at the experimenter's station, the whole stock may be available and could be thoroughly mixed, halved, quartered, etc., until a sufficiently small aliquot is obtained. At a later stage, it might be necessary to use a multi-stage system of sampling, such as choosing by randomization (1) from among all merchants handling the variety, (2) from seed lots of the chosen merchants, (3) from bags within the chosen seed lots and, (4), by sampling the chosen bags.

Sufficient pots would then be filled, each with the same quantity of soil, and the seed sown. We shall suppose here that 20 pots have been found to be sufficient. Uniformity of watering, fertilizing, etc., would be maintained throughout the growth of the plants. Thus, when the time comes to spray the plants, the situation will have been reduced to requiring two samples of pots of wheat which are all from the same population, spraying the plants of one sample so that they now represent the population of sprayed plants, and allowing the others to continue their growth, representing the population of unsprayed plants. To select these samples we should use our random digits but, before doing so, we should consider carefully what forms of variation can affect the experiment. This is because when we fit a mathematical model for analysis, we shall have to account for all sorts of variation, and randomization is crucial in allowing us to make the assumptions necessary to carry out statistical tests. In the present case, because of slight differences in soil environment, slight differences in aerial environment, and differences in growth potential of seeds planted in different pots, we would expect variation from pot to pot; but if there were 20 pots, arranged compactly in five rows of four pots each, we might not expect any positional effects. Then we could consider the 20 pots as a single homogeneous population, and we would use our random digits to select a sample to be sprayed, while the rest remained unsprayed. What we do to the samples is called the *treatment* and, for this experiment, there are two treatments.

(1) Spraying with fungicide, which we will designate a_1.
(2) Not spraying with fungicide, which we will designate a_0.

This use of letters and suffixes is a useful convention, the letter standing for the substance or operation of the treatment, often called the *factor* that is being tested, and the number for what is called the *level* of the factor. In this case there is but one factor (spraying with fungicide) and two levels—absence denoted by 0 and presence denoted by 1. In other experiments there might be several factors, and factors can be at more than two levels.

The treatments must always be dictated by the null hypotheses, and care must be exercised to ensure that those used are really answering the question set. There must be no possible doubt as to what has caused the measured effect. In the present example the null hypothesis was stated as

spraying with fungicide does not affect the yield

and it is fair therefore to have "spraying with fungicide", and all that that involves (such as the water in which the fungicide is dissolved, any wetting agent necessary to make it stick on the leaves, and any disturbance of the plant caused by droplets during the operation) as one treatment, and none of these things as the other. Such a test would be justified in practice by the argument that if the fungicide is to be applied, all these side effects must be present; the practical alternative is to do nothing. However, if there is an effect resulting from this treatment, we cannot claim that the particular chemical in the fungicide affects the yield. That would require a different experiment in which the a_0 treatment involved the same amount of water and wetting agent as the a_1 treatment, but none of the chemical under test.

Many experimenters are disappointed when it is pointed out that their treatments are really confusing two or more factors. It is therefore worth stressing that when deciding on the actual treatments, everything about them must be exactly the same except for the one factor which is being tested. Furthermore, all subsequent operations must be applied to all treatments equally.

Having decided that there should be two treatments, and having 20 pots available, it would seem reasonable to allocate 10 pots to each treatment. It is not essential to have the same number of items for each treatment, but the precision of an estimate is related to the number of items making up the estimate and, to make our comparison, we want both treatments to be equally precise. This is likely to be so if each involves the same number of pots. To allocate the pots to the treatments, they can be numbered 1 to 20 quite arbitrarily, and then a random sample of 10 selected for treatment a_1. Starting nine digits down the first double column of Appendix Table I (p. 225), i.e. at 00, and using the remainder method (see p. 11) to select 10 out of 20, we find a_1 allocated to pots 20, 4, 19, 9, 18, 8, 10, 11, 12 and 14, leaving the rest for a_0 and giving the following layout for the experiment.

1	2	3	4	5
a_0	a_0	a_0	a_1	a_0
6	7	8	9	10
a_0	a_0	a_1	a_1	a_1
11	12	13	14	15
a_1	a_1	a_0	a_1	a_0
16	17	18	19	20
a_0	a_0	a_1	a_1	a_1

All subsequent operations, such as watering, should be as uniform as possible, and no bias should be allowed to creep in. For example, when the grain is ripe, harvesting should proceed systematically along the rows or columns. Thus the randomization which was made at the beginning ensures that each treatment has the same chance of being harvested at any particular time of day. It would be quite wrong to harvest all the a_0s first, and then all the a_1s, as this would introduce another factor into the test. We should then never be sure if the effect we measured was due to spraying with fungicide or due to time of harvest.

CHAPTER THREE

STATISTICAL ANALYSIS

Let us now consider the results obtained from the experiment designed in the previous chapter.

Table 3.1—Yield of wheat from each pot in the experiment designed in Chapter 2

a_0 pots.	Number	1	2	3	5	6	7	13	15	16	17	Total
	Yield (g)	150	143	130	131	123	105	119	144	114	109	1268
a_1 pots.	Number	4	8	9	10	11	12	14	18	19	20	Total
	Yield (g)	182	190	179	163	176	167	158	174	166	154	1709

We want to test the null hypothesis that spraying with the fungicide has no effect on the yield of wheat. If that is so, then the yield of the population represented by the a_0 pots will be the same as that of the population represented by the a_1 pots. We must, then, estimate the yields of these two populations. The values from the two samples certainly look different, but our problem is to condense these arrays of values to produce a few values which are descriptive of the samples and, at the same time, are good estimates of the populations from which the samples were drawn. Clearly if interest was centred solely on the samples, the totals would describe the situation well, but it is not easy to conceive how these could give information about the population when the number of members in the population is unknown. It might be more fruitful to think of how a reasonable man might picture a population. Over the years, a method which has satisfied many people is that called a *frequency diagram* or *frequency distribution*. Since it is rather difficult to see much sense in samples as small as those in the experiment, Table 3.2(*a*) gives the yields of 100 pots of wheat, all of the same variety and all treated in exactly the same way.

Table 3.2(*b*) shows a picture built up by crosses, each cross representing one yield in the stated range. Two features are apparent. Firstly, the number of values per range is largest near the centre of the diagram and decreases as the range moves away from the centre. In other words, the

Table 3.2(a)—Yields of grain (g) from 100 pots

189	215	158	195	185	198	207	186	164	170
154	195	215	206	199	220	184	204	190	197
178	220	170	200	196	190	163	181	239	226
191	228	162	153	216	204	142	189	235	242
194	126	143	217	173	202	156	209	208	206
196	165	238	244	209	162	228	258	147	157
175	175	187	229	215	181	219	182	185	199
159	212	181	172	171	179	187	219	201	204
167	201	209	153	147	167	192	163	172	186
211	198	206	154	164	191	199	187	176	198

Table 3.2(b)—Range of yields and frequency diagram

Range of yields	Number of yields in range	Frequency diagram
120–129	1	x
130–139	0	
140–149	4	xxxx
150–159	8	xxxxxxxx
160–169	9	xxxxxxxxx
170–179	11	xxxxxxxxxxx
180–189	14	xxxxxxxxxxxxxx
190–199	17	xxxxxxxxxxxxxxxxx
200–209	15	xxxxxxxxxxxxxxx
210–219	9	xxxxxxxxx
220–229	6	xxxxxx
230–239	3	xxx
240–249	2	xx
250–259	1	x

population exhibits a central tendency. We could well imagine that the position of this central tendency could distinguish populations. For example, another population might have this area around 250 g instead of around 190 g, and reasonable men would agree that the populations were different for that reason. This definition would be useful also from the practical point of view. Many experiments are made to enable the experimenter to predict what will happen when someone else adopts his recommendation, and the experimenter, having investigated this population, would like to be able to say that plants of this variety of wheat, grown in this way, will produce a definite quantity of grain, say 190 g. Although this is clearly not possible, the next best thing for predictive purposes would be to be able to say that most people would get around 190 g if they grew this variety in this way. Secondly, there is a spread and a general shape to the outline of the crosses in the picture, and we could imagine this spread or shape being different for different populations. Again this would have practical significance, in that the position of a central tendency would be much more meaningful with a narrow spread than with a wide spread.

Location of a population

It now remains to define these two features more precisely. The position of the central tendency, or *location* of the population, which is the term often used by statisticians, may be defined in three ways.

(1) By the *mode* which, if the values are assigned to discrete ranges as in Table 3.2(b), is the range which contains most values: 190–199, in the present case. If the population were shown as a continuous curve, the mode would be the value where the maximum point of the curve occurred.

(2) The location could be defined by the *median*, which is the middle value, when all the values are set out in ascending or descending order of magnitude. In the present example, as there is an even number of values (100), the median would lie between the 50th and 51st values, counting in either ascending or descending order, and would be taken as mid-way between them. In this case, the 50th and 51st values are both 191.

(3) The third definition is the arithmetic *mean*, which is obtained simply by summing all the values and dividing the result by the number of values: in this case $\frac{19067}{100} = 191$ to the nearest gram.

From the practical point of view of trying to find a single value which, when used for prediction, would be one that most people following the recommendation would approach, the best definition depends on the shape of the frequency distribution. In the example in Table 3.2(b) the median and mean are the same, and both occur in the section of the range that formed the mode. However, consider the data (shown in Table 3.3) obtained by trapping mice at several sites in a wood.

Table 3.3—Observations and frequency diagram from a trapping experiment

Number of mice caught per site	Number of sites at which this number was caught	Frequency diagram
0	1	x
1	4	xxxx
2	10	xxxxxxxxxx
3	5	xxxxx
4	4	xxxx
5	4	xxxx
6	3	xxx
7	3	xxx
8	2	xx
9	2	xx
10	2	xx
	Total 40	

Now the mode is 2 mice per site, the median lies between the classes of 3 and 4 mice per site, and the arithmetic mean is $\frac{168}{40} = 4\cdot2$. The best description of the location of this population is more questionable. From a purely descriptive point of view, the mode is useful and clearly understood; unfortunately, mathematical manipulation of it has not proved easy, and no good statistical methods based on the mode have been devised. On the other hand, both the median and the mean can be manipulated and lead to useful methods of estimating population values from sample values. In general, the mean is used when the population is symmetrical, as in Table 3.2(b) where, as we have seen, the median and mode have the same value as the mean. The median is more appropriate for skewed distributions (such as that just considered) since, in such a case, the mean is too near the longer tail of the distribution to represent the majority. This forms a test for skewness of a population; if the mean is greater than the median, there are more members with high values than would be present in a symmetrical distribution, and the distribution is said to be skewed to the right; likewise if the mean is less than the median, there are more members with low values, and the distribution is skewed to the left.

Although in some biological work it is of interest to determine the shape of frequency distribution curves by recording very large samples, in most experiments very small samples are involved, and it is seldom possible to draw any very precise conclusions about the shape of the distribution of the population from the shape exhibited by the sample. Indeed it may be dangerous to do so, as is illustrated by drawing samples of 10 at random from the data in Table 3.2(a). Consider the values in this table to be numbered consecutively down the columns, starting at the top of the extreme left-hand column, and working down each column in turn. Using the random numbers in Appendix Table I and starting at the top of the second column of double digits (06), sampling without replacement gives the values 196, 186, 153, 157, 206, 195, 170, 199, 209 and 189. Using the same divisions of the total range as before gives the frequency diagram shown in Table 3.4.

There are only two values greater than the mode, but five smaller. The median is 192, the mean 186, and it requires a lot of imagination to consider that the diagram is of the same shape as that of the population from which it was derived. If this was the only information available, a skewed distribution would certainly be considered. Thus in experiments involving small samples, it is better to rely on the shape known from experience with similar data from other studies, where larger sample have been employed.

Table 3.4—Range of yields and frequency diagram of a random sample drawn from the data of Table 3.2(a)

Range of yields	Number of yields in range	Frequency diagram
150–159	2	xx
160–169	0	
170–179	1	x
180–189	2	xx
190–199	3	xxx
200–209	2	xx

It is perhaps fortunate for experimental biology that most measurements or weights of plants and animals follow reasonably closely a symmetrical distribution well known to statisticians as the *normal distribution* (or Gaussian distribution). There is a rather remarkable theorem, known as the Central Limit Theorem, which states, approximately, that with very few exceptions the averages of random samples drawn from any distribution, no matter what shape, will be approximately normally distributed. Biological data are usually expressed as some form of average, e.g. yield of wheat per plant comes from the average weight of a number of grains. Even when a single item like length of stem is considered, it will have resulted from the average of a large number of genetic and environmental disturbances. The normal distribution is illustrated in figure 3.1. In this model the mean, median and mode coincide. It can be seen that it would reasonably fit the data in Table 3.2.

There are three aspects to all statistical analyses. There are the samples which are actually recorded: the data from these are real. Then there is the population from which we believe the sample was drawn, and we conceive that this population has certain *true* values, though we can never measure them directly. To connect the two we use the mathematical model. We know which manipulations to apply to a sample of random variates to obtain estimates of the mathematical population, and we apply these to our sample data to obtain estimates of the true values that we are seeking for our biological population.

Here, then, we would consider that the mean was a good measure of location and, if we could estimate the true population means from our sample values, and they differed, it would make sound biological sense to say that spraying with fungicide affected the yield of this variety of wheat. Thus we require unbiased estimates of the population means.

In the mathematics of random variables, an estimate T of a statistic θ is said to be unbiased if the average value of T from its whole distribution

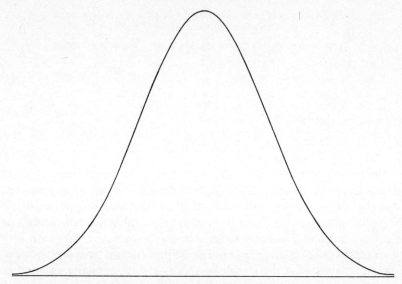

Figure 3.1—Standard normal distribution.

(usually written $E(T)$) is equal to θ. This is usually expressed as $E(T) = \theta$. It can be shown that in any distribution for which a mean exists, the mean of a random sample is an unbiased estimate of the mean of the population. This is obvious without mathematical formality, since all possible samples will contain all members of the population in equal numbers; therefore the mean of these means must be the mean of the population. Since we considered that we had drawn our samples at random from our conceptual populations, we would be prepared to take the means of our samples as unbiased estimates of the means of these populations.

Dispersion of a population

Like the location of the central tendency, the *spread* or *dispersion*, as it is known to statisticians, can be estimated in a number of ways. One way is to consider the *range*, i.e. the difference between the highest and lowest values. It is possible to make some population estimates from samples in this way but, clearly, very little of the information provided by the sample would be used. Only the two extreme values would be used and the intermediate values would contribute nothing. We would not, therefore, expect the range to be a very efficient estimate. When the mean is a good estimate of the location of a distribution, it is reasonable to think that the spread of values about the mean would be a good indication of the dispersion. We can easily calculate the difference between each value and the mean of the sample, which in statistical language is known as the *deviation* from the mean, or *deviate*.

For algebraic purposes, an array of values is considered as

$$x_1, x_2, x_3, \ldots x_i \ldots x_n$$

where x_i, the ith value, is taken to indicate the general term and x_n is the last value, so that n is the number of values. The total of the array is written as

$$\sum_{i=1}^{n} x_i \quad \text{and the mean as} \quad \frac{1}{n} \sum_{i=1}^{n} x_i.$$

Provided there is no possibility of confusion about the xs with which we are dealing, the total may be simplified to $\sum x$, meaning the sum of all the xs under consideration, and the mean may be written $\bar{x}$. Thus the general deviate may be written $x_i - \bar{x}$, and it is immediately apparent that if we sum the deviates to find the total spread we obtain

$$\sum_{i=1}^{n} (x_i - \bar{x}) = \sum_{i=1}^{n} x_i - n\bar{x}$$

(because $\bar{x}$ is constant for every term of the summation, and therefore has to be subtracted n times)

$$= \sum_{i=1}^{n} x_i - n \left(\frac{\sum_{i=1}^{n} x_i}{n} \right)$$
$$= 0$$

As this would be true for all samples, the method would be unlikely to allow the drawing of comparisons. If, however, the deviates were squared before adding, we should obtain a value which would be zero if all the values in the samples were the same, i.e. if there was no variation; otherwise the value would be positive and become larger as the variation increased. This operation produces a value known as *the sum of squares of the deviations from the mean*, often abbreviated to *sum of squares* and given the symbol S.S. The method leads to easy estimation of one of the parameters determining the mathematician's normal distribution. Perhaps the point to emphasize is that, unlike some branches of mathematics where there is a correct way of doing something and all other ways are wrong, in statistics we often have to decide which of several possible methods to use: none can be said to be either absolutely correct, or even the best that will ever be discovered. This is true in the present case, and all that can be said in favour of the method is that it has been found to lead to consistent results, and has been found more generally useful in biology than has any other method.

We could, therefore, estimate the spread of each of our samples as $\sum(x-\bar{x})^2$, where x now applies to all members of that sample. But this is rather a tedious process and a little algebraic simplification might help. In normal algebra,

$$\sum(x-\bar{x})^2 = \sum(x^2 - 2\bar{x}x + \bar{x}^2)$$

In summation the order does not matter, i.e. in the expression $\sum(a+b) = \sum a + \sum b$, the same answer is obtained by adding all the as first, then all the bs, and adding the two totals together, as by adding each a to b, and then adding the $(a+b)$s together. It is also important to note that multiplication of each value by a constant before the addition is the same as multiplication of the sum of the values by the same constant. We have already seen that adding a constant n times produces n times the constant, and so

$$\sum(x^2 - 2\bar{x}x + \bar{x}^2) = \sum x^2 - 2\bar{x}\sum x + n\bar{x}^2$$

$$= \sum x^2 - \frac{2\sum x \sum x}{n} + n\left(\frac{\sum x}{n}\right)^2$$

$$= \sum x^2 - \frac{(\sum x)^2}{n}$$

giving a very simple and easily used expression which, in words, is the sum of the squares of the values minus the total, squared and divided by the number of values. Because this calculation is done in two parts, the separate parts are often given names. $\sum x^2$ is called the *crude sum of squares*, and $(\sum x)^2/n$ the *correction factor* (C.F.). There may be various explanations for the origin of the second name. The most likely is that when this system was first used the $\sum x^2$ term was thought to be giving the sum of the squares of the deviations from zero and, to find the sum of squares of the deviations from the mean, it was necessary to correct this value by subtracting n times the square of the mean ($n\bar{x}^2 = (\sum x)^2/n$, as we have seen). Whatever the explanation, it is important to realize that the correction factor is not some magical device for correcting the calculations, as the name might imply, but is simply the name given to one part of a very simple algebraic function.

STATISTICAL ANALYSIS

There are ways of easing the burden of arithmetic even more. The first is a special case when there are only two members in the sample, e.g. x_1 and x_2.

Then $\sum x^2 - \dfrac{(\sum x)^2}{n}$ becomes $x_1^2 + x_2^2 - \dfrac{(x_1 + x_2)^2}{2}$

$$= \tfrac{1}{2}(2x_1^2 + 2x_2^2 - x_1^2 - 2x_1 x_2 - x_2^2)$$
$$= \tfrac{1}{2}(x_1^2 + x_2^2 - 2x_1 x_2)$$
$$= \tfrac{1}{2}(x_1 - x_2)^2$$

← S.S. $n=2$

This is very useful in experimentation; in words, the sum of squares for two members, often called the *sum of squares of a difference*, is half the square of the difference.

Another useful device is called *coding the data*. One method is to subtract a constant. The effect of this is to reduce the mean by the same amount, but not to affect the sum of squares at all. Thus if we have our usual array of xs and subtract from them a constant c, they become

Coding

$$(x_1 - c), (x_2 - c), (x_3 - c), \ldots (x_i - c), \ldots (x_n - c)$$

The total is $\sum\limits_{i=1}^{n}(x_i - c) = \sum\limits_{i=1}^{n} x_i - nc$

and so the mean is

$$\dfrac{\sum\limits_{i=1}^{n} x_i - nc}{n} = \dfrac{\sum\limits_{i=1}^{n} x}{n} - c = \bar{x} - c$$

and the S.S. in the form of $\sum(x - \bar{x})^2$ becomes

$$\sum\{(x-c)-(\bar{x}-c)\}^2 = \sum(x - \bar{x})^2 \quad \text{as before.}$$

Another form of coding is to multiply all the data by a constant. This is a useful way to remove decimals and fractions, e.g. if we have $x = 12\tfrac{1}{4}, 13\tfrac{1}{2}, 12\tfrac{7}{8}$, we can multiply by 8 to get 98, 108, 103, which makes squaring much simpler. When coding by multiplying by a constant, the real mean is obtained by dividing the coded mean by the same constant, and the real S.S. by dividing the coded sum of squares by the square of the constant. Thus the array becomes $cx_1, cx_2, cx_3, \ldots cx_n$.

$$\text{Total} = \sum(cx)$$

$$\text{Mean} = \frac{\sum(cx)}{n} = \frac{c\sum x}{n} = c\bar{x}$$

$$\text{S.S.} = \sum(cx - c\bar{x})^2 = \sum\{c(x-\bar{x})\}^2 = c^2\sum(x-\bar{x})^2$$

Returning to the data at the beginning of this chapter, we could subtract 100 from each value, and the sum of squares of the sample receiving the a_0 treatment will be

$$50^2 + 43^2 + 30^2 + 31^2 + 23^2 + 5^2 + 19^2 + 44^2 + 14^2 + 9^2 - \frac{268^2}{10}$$

$$= 9338 - 7182 \cdot 4 = 2155 \cdot 6$$

For the sample receiving the a_1 treatment, the corresponding sum will be

$$82^2 + 90^2 + 79^2 + 63^2 + 76^2 + 67^2 + 58^2 + 74^2 + 66^2 + 54^2 - \frac{709^2}{10}$$

$$= 51411 - 50268 \cdot 1 = 1142 \cdot 9$$

However, we require a measure of the spread of the population and, clearly, the sum of squares of a sample will depend on how many members there are in the sample; it will therefore be of little use by itself for estimating the population sum of squares, when the number of members in the population is unknown. When considering the location, we overcame a similar problem by dividing the total by the number of observations, thus producing the mean. This was quite independent of the number of members in the sample, and so proved a suitable value to describe a population of unknown number. Similarly, we could expect to calculate a mean spread of a sample, i.e. an average variation. To do this, the sum of squares must be divided by something. At first thought we might consider dividing by the number of deviations that were squared, which is the same as the number of values in the sample. However, all the deviates in the sample are not independent of one another. We have seen that $\sum(x-\bar{x}) = 0$, i.e.

$$(x_1 - \bar{x}) + (x_2 - \bar{x}) + (x_3 - \bar{x}) + \ldots + (x_n - \bar{x}) = 0$$

Therefore

$$(x_1 - \bar{x}) = -\{(x_2 - \bar{x}) + (x_3 - \bar{x}) + \ldots + (x_n - \bar{x})\}$$

Thus one of the deviates can always be calculated by knowing all the others. Moreover, as we do not know the true mean, our deviations are not true deviations. It can be shown easily that sums of squares of deviations are at a minimum when they are deviations from the mean of the actual values in the sample.

(If we write

$$\sum_{}^{n}(x-\mu)^2$$

for the deviations from any fixed value, and differentiate with respect to μ we have

$$-2\sum_{}^{n}(x-\mu).$$

The sum of squares will be at a minimum when

$$-2\sum_{}^{n}(x-\mu) = 0,$$

i.e. when

$$\sum x = n\mu \quad \text{or} \quad \mu = \frac{\sum x}{n} = \bar{x}.)$$

Thus the average variation, calculated from the total variation of our sample by dividing by the number of observations in the sample, would be an underestimate of the population variation.

It turns out that the unbiased estimate of the average spread, or *variance* of a population of random variates, to give it its mathematical name, is obtained by dividing the sample sum of squares by the number of independent deviations, i.e. by one less than the number of observations in the sample. This measure is called the *mean square* and is given the symbol M.S. Thus

$$\text{M.S.} = \frac{\text{S.S.}}{n-1}.$$

That this is an unbiased estimate of the variance will be justified when other manipulations of mean squares have been established. The denominator $(n-1)$ is also given a name in statistics. It is known as the *degrees of freedom* and abbreviated D.F. Hence

$$\text{M.S.} = \frac{\text{S.S.}}{\text{D.F.}}.$$

More generally, degrees of freedom can be defined as the number of observations less the number of parameters fixed by those observations; in the present case one parameter, the mean, has been fixed by the observations.

In this example, there are 10 observations in each sample, and so 9 D.F. in each case. The mean square for the a_0 sample is

$$\frac{2155 \cdot 6}{9} = 239 \cdot 51,$$

and for the a_1 sample is

$$\frac{1142 \cdot 9}{9} = 126 \cdot 99.$$

However, these samples were originally drawn by the process of randomization from the same population of wheat plants, and there is no reason to believe that application of the fungicide would alter the spread of the population. We would claim, therefore, that these two mean squares are both estimates of the same variance. The best estimate of that variance would be derived from a combination of the two. In this case we could simply take the mean of the two mean squares but, more generally, they should be weighted according to the information each contains, i.e. according to their degrees of freedom. To do this, we add the sums of squares and divide by the total degrees of freedom. In this case, then, the variance is estimated as

$$\frac{2155 \cdot 6 + 1142 \cdot 9}{9+9} = \frac{3298 \cdot 5}{18} = 183 \cdot 25.$$

At first glance, the above argument may seem doubtful, since the mean square for a_0 is nearly twice that for a_1. They look very different but, as when judging the shape of the distribution from a small sample, it must be remembered that these are estimates which are themselves subject to sampling variation. They cannot be expected to be exactly the same, even when drawn from the same population. Therefore we should require large differences between mean squares to convince us that our treatment had affected the plant's natural variation. A test for mean squares is given later. If it were applied, it would be seen that with samples as small as this, one mean square would have to be four times the other before we suspected that our assumption of equal variance was not true.

We have now defined the samples and estimated the locations and dispersion of the populations from which they were drawn; but the two measures are not in the same units: the means are in grams and the mean square is in grams^2. In the mathematics of distributions, the variance is usually denoted by σ^2 and the square root of this (σ) is termed the *standard deviation*, i.e. it is a sort of mean deviation, expressed in the same units as the original variable. The same operation can be carried out on mean squares (often denoted by s^2) to produce s, an estimate of the standard deviation and referred to as the *standard error of a single observation*. In the present case

$$s = \sqrt{183 \cdot 25} = 13 \cdot 54.$$

In this way, the populations can be described by the estimates of the means, and by a value which indicates how much the individuals vary around the mean, expressed in the same units.

Standard error of means

The objective of the experiment requires a comparison of the two true means, and it will be clear that the estimates of the means are themselves subject to sampling variation. We must therefore consider a population of means, and try to estimate how means drawn from the same population would vary around their mean. Intuitively, we would expect means to vary less than individuals, e.g. consider the numbers 0, 2, 4, with a mean of 2, and take means of all possible pairs, giving 1, 2, 3, with clearly less variation around the mean of 2. It would, of course, be possible to repeat the experiment many times, calculate the means, their sums of squares and mean squares but this would involve a lot of work. It would therefore be better to see if the variation of the means could be estimated from the variation of individuals.

It is simplest to start by considering the variation of a total of two values selected at random from the same population. Let us denote them as

$$x_1 \quad x_2 \quad x_3 \ldots x_n \quad \text{for the first of each pair.}$$

$$y_1 \quad y_2 \quad y_3 \ldots y_n \quad \text{for the second.}$$

Now if we add the pairs of values together to obtain totals of two, we have the following array:

$$(x_1 + y_1), (x_2 + y_2), (x_3 + y_3), \ldots, (x_n + y_n)$$

and

$$\text{M.S. of } (x+y) = \frac{\sum[(x+y) - \overline{(x+y)}]^2}{n-1}$$

But

$$\overline{(x+y)} = \frac{\sum(x+y)}{n} = \frac{\sum x}{n} + \frac{\sum y}{n} = \bar{x} + \bar{y}$$

Therefore

$$\text{M.S. of } (x+y) = \frac{1}{n-1}\sum(x+y-\bar{x}-\bar{y})^2$$

$$= \frac{1}{n-1}\sum[(x-\bar{x})+(y-\bar{y})]^2$$

$$= \frac{1}{n-1}\sum[(x-\bar{x})^2 + 2(x-\bar{x})(y-\bar{y}) + (y-\bar{y})^2]$$

$$= \frac{\sum(x-\bar{x})^2}{n-1} + \frac{2\sum(x-\bar{x})(y-\bar{y})}{n-1} + \frac{\sum(y-\bar{y})^2}{n-1}$$

$$= \text{M.S.}_{\cdot x} + \frac{2\sum(x-\bar{x})(y-\bar{y})}{n-1} + \text{M.S.}_{\cdot y}$$

Now consider the middle term. Since $\bar{x}$ is the mean of all the xs, it is obvious that about half the xs will be greater than $\bar{x}$ and half will be less than $\bar{x}$. Then the values of $x-\bar{x}$ will be about half positive and half negative. The values of $y-\bar{y}$ will be half positive and half negative, and when we multiply $(x-\bar{x})$ by $(y-\bar{y})$ we obtain

$$+ \times +, \; + \times -, \; - \times + \; \text{and} \; - \times - \quad \text{or} \quad + \; - \; - \; +$$

If there is no relation between the x and y, a positive $(x-\bar{x})$ will appear with a positive $(y-\bar{y})$ just as often as it will appear with a negative $(y-\bar{y})$, and likewise for a negative $(x-\bar{x})$; then there will be approximately equal numbers of positive and negative products, and so in the long run the sum of these products will tend to zero. Then

$$\text{M.S.}_{\cdot(x+y)} \simeq \text{M.S.}_{\cdot x} + \text{M.S.}_{\cdot y}$$

i.e. the mean square of the sum of two sets of values = the sum of the mean squares. The argument easily extends to totals of any number of values when there is no association between them.

$$\text{M.S.}_{(x_1+x_2+x_3\ldots+x_r)} = \text{M.S.}_{\cdot x_1} + \text{M.S.}_{\cdot x_2} + \text{M.S.}_{\cdot x_3} \ldots + \text{M.S.}_{\cdot x_r}$$

In an experiment where there is but one M.S.$_{\cdot x}$, applicable to each and every observation

$$M.S._{x_1} = M.S._{x_2} = M.S._{x_3} \ldots = M.S._{x_r}$$

so

$$M.S._{\Sigma x} = r\, M.S._x$$

or, in words, the variance of a total of r independent values, all from the same population, can be estimated as r times the estimate of the variance of the population.

However, we are seeking an estimate of the variance of means

$$M.S._{\bar{x}} = M.S._{(\Sigma x)/r}$$

for a mean of r values. We have already seen that if each value is multiplied by a constant c

$$S.S._{cx} = c^2\, S.S._x$$

Since

$$M.S. = \frac{S.S.}{D.F.}$$

$$M.S._{cx} = c^2\, M.S._x$$

Since converting the M.S. for a total of r values into a M.S. for a mean of r values is equivalent to dividing each value by r before squaring and adding, then

$$M.S._{(\Sigma x)/r} = \left(\frac{1}{r}\right)^2 M.S._{\Sigma x} = \frac{r\, M.S._x}{r^2} = \frac{M.S._x}{r}$$

In words, the estimate of the variance of a mean of r independent values, drawn from the same population, is the estimate of the variance of the population divided by r.

Since the standard deviation is always defined as the square root of the variance, the standard error of a mean of r values will be $\sqrt{(M.S._x/r)}$. In our example, the S.E. of the means will therefore be

$$\sqrt{\frac{183\cdot25}{10}} = \pm 4\cdot28.$$

Knowing how to estimate the variance of a mean enables us to appreciate better why the mean square is an unbiased estimate of the variance. Let us insert the population mean μ into the expression for the sum of squares of a sample, using again the operator $E(\ )$, representing the average over a long run of samples.

$$E[\sum(x-\bar{x})^2] = E[\sum\{(x-\mu)-(\bar{x}-\mu)\}^2]$$
$$= E[\sum\{(x-\mu)^2 - 2(\bar{x}-\mu)(x-\mu) + (\bar{x}-\mu)^2\}]$$
$$= E[\sum(x-\mu)^2 - 2(\bar{x}-\mu)\sum(x-\mu) + n(\bar{x}-\mu)^2],$$

since $\bar{x}$ and μ are constant over the n terms of the summation.

Now $\sum(x-\mu)$ over n terms $= \sum x - n\mu = n(\bar{x}-\mu)$. Therefore

$$E[\sum(x-\bar{x})^2] = E[\sum(x-\mu)^2 - n(\bar{x}-\mu)^2]$$

By definition $E[(x-\mu)^2]$ is the variance of $x = \sigma^2$. Therefore

$$E[\sum(x-\mu)^2] \text{ over } n \text{ terms} = n\sigma^2$$

and $E[(\bar{x}-\mu)^2]$ is, again by definition, the variance of $\bar{x} = \sigma^2/n$. Therefore

$$E[\sum(x-\bar{x})^2] = n\sigma^2 - \frac{n\sigma^2}{n} = (n-1)\sigma^2$$

Therefore

$$E\left[\frac{\sum(x-\bar{x})^2}{n-1}\right] = \sigma^2$$

Thus the mean square $\left(\dfrac{\text{sum of squares}}{\text{degrees of freedom}}\right)$ is an unbiased estimate of the variance.

Standard error of a difference

One further manipulation of these estimates is required before a test of difference between populations can be made. If the means are subject to sampling variation, the difference between them will be also, and we shall require an estimate of the variance of differences between two means. Following the argument we used to obtain the M.S. for totals, we could consider an array of values of $x-y$ where both x and y are randomly chosen from a normal population. Following through the algebra on pp. 29–30, the final line becomes

$$\text{M.S.}_{(x-y)} = \text{M.S.}_x - \frac{2\sum(x-\bar{x})(y-\bar{y})}{n-1} + \text{M.S.}_y$$

By the same reasoning, the middle term becomes zero in the long run,

and so
$$\text{M.S.}_{\cdot(x-y)} = \text{M.S.}_{\cdot x} + \text{M.S.}_{\cdot y}$$

Likewise the estimates of the variance of the difference between two means of r values, drawn from the same population, will be

$$\frac{\text{M.S.}_{\cdot x}}{r} + \frac{\text{M.S.}_{\cdot x}}{r} = \frac{2\text{M.S.}_{\cdot x}}{r}$$

and its standard error will be $\sqrt{(2\text{M.S.}_{\cdot x}/r)}$ or $s\sqrt{(2/r)}$ using the symbol defined earlier. If the samples are of different sizes (say r_1 and r_2), the standard error of the difference will be

$$s\sqrt{\left(\frac{1}{r_1} + \frac{1}{r_2}\right)}.$$

In the present example, both sample means are derived from 10 values, and so the S.E. of the difference between them will be

$$\sqrt{\frac{2 \times 183 \cdot 25}{10}} = \pm 6 \cdot 05.$$

Coefficient of variation

There is one other measure of variation that should be mentioned. All those described so far depend on the units in which the original samples were measured. Admittedly it is easy to change units, because if c is a conversion factor from one set of units to another, we have seen that the mean of cx is c times the mean of x. Also

$$\text{M.S.}_{\cdot cx} = c^2 \text{M.S.}_{\cdot x}$$

and it follows that

$$\text{S.E.}_{\cdot cx} = c\text{S.E.}_{\cdot x}$$

Nevertheless, it is desirable in biological experimentation to be able to compare the relative variabilities of populations or, more particularly, the variability of different experiments. Such a unit-free measure is called the *coefficient of variation* (C.V.), which is simply the standard error of a single observation, expressed as a percentage of the mean, i.e. $(s \times 100)/\bar{x}$. In the wheat example,

$$\text{C.V.} = \frac{13 \cdot 54 \times 100}{148 \cdot 85} = 9 \cdot 1\%$$

STATISTICAL ANALYSIS

The coefficient of variation has no use in testing or estimating, but provides a good yard-stick for appreciating the precision possible in an experiment, thereby aiding decisions on the size of sample to be taken, and allowing comparisons of variability between experiments.

CHAPTER FOUR

TESTING AND SETTING CONFIDENCE LIMITS

We have argued that in biological experiments we can seldom test hypotheses by comparing whole populations, when, of course, if we had the hypothesis that the means of two populations were the same, we could calculate them exactly and any difference in the two values would disprove the hypothesis. Instead, we have the situation of estimating the means of the populations by measuring samples. Because of natural variation we know full well that our samples will not give an exact measure of the population value, and that if we took another sample we should get a different answer. However, we have also seen that these sample values follow a distribution with a tendency to cluster around the true population value. So in testing we base our argument on this distribution of values.

The mathematical normal distribution is defined by a simple expression

$$y = \frac{1}{\sigma \sqrt{(2\pi)}} e^{-\frac{1}{2}(x-\mu)^2/\sigma^2}$$

where x and y are the usual coordinates, μ is the arithmetic mean, σ^2 is the variance $\sum (x-\mu)^2/n$, and σ therefore the standard deviation. Thus only two parameters are required, μ and σ, and these, perhaps fortunately, are two things that we thought worth estimating for practical reasons. We have seen from the coding argument that we can add or subtract a quantity to each value without altering the variance, but simply altering the mean by the same amount; and we can multiply or divide by a constant when we alter the mean likewise and alter the variance by the square of the constant.

If we have a variate x which is normally distributed with mean μ and variance σ^2, $x(N, \mu, \sigma^2)$ as we write it, then $x - \mu$ is $(N, 0, \sigma^2)$ and $(x-\mu)/\sigma$ is $(N, 0, 1)$.

All variates which are normally distributed can be brought to a common normal distribution by subtracting the mean and dividing by the standard deviation. They are then known as standard normal deviates and extensive tables of frequencies have been published (e.g. Fisher and Yates, 1963).

From these tables we can determine how frequently various groups of values of $(x-\mu)/\sigma$ can occur. For example, 68 per cent of the values lie between -1 and $+1$ and only 5 per cent of the values are either less than -1.96 or more than $+1.96$. Therefore, if we knew that our population had a mean μ and standard deviation σ, and we picked up an individual x such that the value of $(x-\mu)/\sigma$ was more than 1.96 or less than -1.96, we would know that the frequency with which we could pick such a value is less than 5 per cent.

In statistics the *probability* of a happening is defined as the relative frequency with which it can happen. If we know that the frequency of picking such a value from our population is only 5 per cent, then we accept that the probability that we will pick such a value is also 5 per cent, though in probability terms it is more usual to use a scale of 0 to 1 instead of 0 to 100 when a probability of 5 per cent becomes $P = 0.05$.

If we had this known population and picked up this supposed member, we are then faced with a decision: if it is a member of the population we had a very small chance of picking it, so the alternative that it is not really a member of that population at all is very attractive.

Furthermore it is possible to determine mathematically the frequency distribution of a similar statistic, namely $(x-\mu)/s$ where s is the S.E. of x or estimate of the standard deviation calculated from a sample. This frequency distribution was worked out by a head-brewer of Guinness who always signed his papers "Student" and so is known as Student's t-distribution. It is similar to the normal distribution, but its shape alters as the size of the sample used to calculate s alters.

Appendix Table II (p. 226) gives the values of t for cutting off the outside 5 per cent, 1 per cent and 0.1 per cent of the distribution. It is seen that the size of t varies as the D.F., which is the D.F. for the s in the formula. Degrees of freedom measure the amount of information, and it is reasonable that the more information there is, the closer will t get to the normal distribution value; when D.F. are infinite $t = 1.96$ for 5 per cent as expected.

We want to use these tables to see if there is any difference between the means of two populations. We first make a null hypothesis that there is no *true* difference, i.e. μ_0 for $a_0 = \mu_1$ for a_1.

Now the sample means estimate the population means; if the hypothesis is true, then $\bar{a}_1 - \bar{a}_0$ is an estimate of $\mu_1 - \mu_0$ which we say should be zero.

$$\frac{(\bar{a}_1 - \bar{a}_0) - (\mu_1 - \mu_0)}{\text{S.E. of } (\bar{a}_1 - \bar{a}_0)}$$

will be distributed as t with the D.F. of the S.E.

TESTING AND SETTING CONFIDENCE LIMITS

Hence if the hypothesis is correct and $\mu_1 - \mu_0 = 0$

$$\frac{\bar{a}_1 - \bar{a}_0}{s\sqrt{\frac{2}{r}}}$$

is distributed as t. In our example $\bar{a}_1 - \bar{a}_0 = 170 \cdot 9 - 126 \cdot 8 = 44 \cdot 1$ (from p. 17) and

$$s\sqrt{\frac{2}{r}} = 6 \cdot 05 \text{ (p. 33)} \quad \text{so} \quad t = \frac{44 \cdot 1}{6 \cdot 05} = 7 \cdot 29$$

with 18 D.F. We see that this is nearly twice the value for $P = 0 \cdot 001$ of $3 \cdot 92$ in Appendix Table II. We can therefore say that this difference (or more) between two sets of 10 observations could have arisen less than once in a 1000 times due to chance if the treatments did not differ in their effects. This is the only sort of precise statement that can be made about variable material, so some system which would be applicable to all experiments is desirable.

It was stated earlier that no experiment can verify a hypothesis completely; it can only give evidence for or against it. The use of statistical methods can take this a stage further and give a measure of the degree of certainty of the evidence. To do this we proposed the null hypothesis that spraying the fungicide will *not* affect the yield. Having made the experiment it is seen that if the null hypothesis were true, our result could have been got less than once in a 1000 times, a very unlikely occurrence. We are then faced with a decision. Do we believe that this very unlikely event has occurred, or do we prefer the alternative which is that the null hypothesis is not true? If the null hypothesis is not true, then the original hypothesis is acceptable, because it is simply the reverse of the null hypothesis. In experimentation we speak of the result as being *significant* at the appropriate level of probability, in our case significant at the 0·1 per cent level of probability or significant ($P < 0 \cdot 001$). In real life we have to decide what level of significance we will accept as indicating a true effect. This is not a statistical decision and cannot be put in any formal terms. It depends upon the reason for the experiment, upon the experimenter's personal caution, and upon the reasonableness of the hypothesis. For instance, if the experiment was testing a drug which either cured or killed a patient, most experimenters would require a very high significance in favour of a cure before they would recommend its use; the 1 in 1000 level would not satisfy many, whereas if the experiment was one of a series where confirmation of the hypothesis was going to lead to further experiments to widen the hypothesis, a probability of 0·1 might well be acceptable because, if it had been a chance result, the experimenter would continue to do some unnecessary work, whereas if it was a true effect and he discarded it, the whole of the future programme might be adversely affected. However, the most commonly used values of acceptance are the 5 per cent level ($P < 0.05$) which is referred to simply as *significant*, the 1 per cent ($P < 0.01$) which is referred to as *highly significant*, and 0·1 per cent ($P < 0.001$) which is referred to as *very highly significant*. It is a convention in most journals to refer results to these levels of significance by placing one star over the difference if it is significant at the 5 per cent level, two stars for the 1 per cent level and three stars for the 0·1 per cent level, so that our result might be written:

$$\text{effect of spraying fungicide} = +\overset{***}{44\cdot1}\text{ g/pot}$$

As the knowledge of statistics spreads, it is much better to report the difference, the standard error of the difference, and the degrees of freedom, i.e.

$$44\cdot1 \pm 6\cdot05 \text{ (18 D.F.)}$$

The reader can then work out the value of t for himself and make his own decision as to the acceptability of the result. Using the word *significant*, or using stars, invites the reader to draw a sharp line between results which are and those which are not significant, leading to the thought that, if it is significant, the result shows a very true and certain effect, whereas if it is not there is no effect at all. Having learnt what a significance test is about, it is seen that this is not the case at all. The effect can be true or not at any level of probability; it is just that the evidence for it being true is slightly greater when t has the 5 per cent value than when it has the 5·1 per cent value. Thus fussing with exact levels of probability in biological experimentation is seldom fruitful. Statistical tests give confidence in biological interpretation and help to pick out the important aspects of a problem. To use them otherwise is to strain at the gnat of exactitude of the mathematics but to swallow the camel of personal belief in what is a rare event. Using the method of giving the difference and its standard error does not cause most experienced readers any great labour in calculation. With use we can soon get a very good idea of what values of t are appropriate to various probabilities, the simple ones being that for S.E.s with 20 or more D.F. t approximates to 2 at 5 per cent and to 3 at 1 per cent, and so we look at the difference and say: it is more than twice its standard error, so the probability of a chance effect of this magnitude must be less than 5 per cent, and so on.

Least significant difference

There are two other ways of using this statistic t. It follows from the t-test that if

$$t = \frac{\text{difference}}{\text{S.E. of difference}}$$

and for a given level of significance $t \geqslant t_s$ where t_s is the value in the table for the appropriate degrees of freedom, then

$$\frac{\text{difference}}{\text{S.E. of difference}} \geqslant t_s$$

for significance at that level, or

$$\text{difference} \geqslant t_s \times \text{S.E. of difference},$$

so many authors work out the value of $t_s \times$ S.E. of the difference and call it the *least significant difference* or L.S.D. and give their results in the tabular form of

	a_0	a_1	L.S.D. $(P = 0.05)$
Yield of wheat (g/pot)	126·8	170·9	12·71

The L.S.D. in this case is t for 18 D.F. at the 5 per cent level (2·10 from Appendix Table II) multiplied by 6·05 (the S.E. of the difference between the two means). This practice ought to be used with care; it is all right when only one null hypothesis can be tested in the experiment, but it may mislead if several possible effects can be got from the same table. Thus supposing we have

0	A	B	L.S.D. $(P = 0.05)$
50·0	73·6	75·8	24·3

the ordinary reader using non-statistical knowledge says B had a significant effect, A did not, therefore B is obviously better than A, whereas the statistical test may have had nothing to do with that comparison and such a conclusion may be quite misleading.

In such cases it is better to give the means and their standard error. This enables the reader to make tests, but the extra effort of obtaining the differences he is interested in and multiplying the S.E. by $\sqrt{2}$ may discourage him from making tests before reading the text to find out what null hypotheses were involved. This method of displaying the results brings up the question of how many significant figures should be used. Generally the greater the variability the smaller is the number of significant figures with any real meaning. There is no single rigorous rule that can be applied to all situations, but a good working rule is: round the original data to three significant figures unless a very low coefficient of variation (less than 4 per cent) is expected, when four significant figures should be retained if available. Then retain all figures when squaring these values or combinations of them, e.g. if the data have two places of decimals the S.S. and M.S. should have four. Then means for tables of results should be rounded to one tenth of their standard errors, and the standard errors given to one more place of decimals than the means themselves.

In our example, the treatment means had a standard error of 4·28, one tenth of which is 0·428, so we would round our means to one place of decimals and set out the results as follows:

	Treatments		
	a_0	a_1	S.E.
Yield of wheat (g/pot)	126·8	170·9	4·28

This rule should always be tempered with biological sense. The number of significant figures should be reduced if those suggested for statistical reasons give a sense of accuracy which cannot be sustained biologically. In this case, it is reasonable to think in terms of 0·1 g of grain from a pot of wheat, but if the S.E. suggested that we rounded yields to 0·1 kg/ha it is ridiculous to think that we would wish to distinguish as little as 100 g from areas which normally yield something like 5000 kg; thus it would make more biological sense to round to the nearest 10 kg. On the other hand we should never display false accuracy simply because the last figure has some biological meaning.

Confidence limits

The remaining use of t is in estimating. Suppose we had drawn our pots of wheat at random from a large population of pots, and we wanted to make a statement about the increase in yield of the whole population if it was sprayed with fungicide. Now the best estimate we have of the difference between the means of the populations is the difference between the means of our samples, 44·1 g/pot. We expect the difference between the means of the populations to be somewhere near that value, but we would be happier if we could say how near that value and say how confident we were of the statement. We can do this by working out what

TESTING AND SETTING CONFIDENCE LIMITS 41

are known as *confidence limits* for this value. These confidence limits are simply $44.1 \pm t_s \times$ S.E. where t_s is the value given in the table for the particular probability chosen. As with testing we can appeal to the *t*-distribution. It has already been said (p. 36) that

$$\frac{(\bar{a}_1 - \bar{a}_0) - (\mu_1 - \mu_0)}{\text{S.E. of } (\bar{a}_1 - \bar{a}_0)}$$

is distributed as t, so putting $(\bar{a}_1 - \bar{a}_0) = d$, the difference measured by our samples and $(\mu_1 - \mu_0) = \delta$, the true difference, we can write

$$\frac{d - \delta}{\text{S.E. of } d} = t$$

for all t defined by the D.F. of the S.E. of d. Tables of t (Appendix Table II) give the largest values of t which can occur with a given probability, e.g. for 18 D.F. and $P = 0.05$, $t = 2.10$, meaning that 5 per cent of possible *t*-values lie outside the range of $t = -2.10$ to $t = +2.10$. Therefore 95 per cent of possible *t*-values lie within $t = -2.10$ to $t = +2.10$, or put more succinctly there is 95 per cent probability ($P = 0.95$) that $-2.10 \leqslant t \leqslant +2.10$.

Therefore from the equation relating t to the differences, there is 95 per cent probability that

$$-2.10 \leqslant \frac{d - \delta}{\text{S.E. of } d} \leqslant +2.10$$

$\therefore \qquad -2.10 \times \text{S.E. of } d \leqslant d - \delta \leqslant +2.10 \times \text{S.E. of } d$

or $\quad d - 2.10 \times \text{S.E. of } d \leqslant \delta \leqslant d + 2.10 \times \text{S.E. of } d \quad$ when $\quad P = 0.95$

The extremes of the range are then referred to as the 95 per cent confidence limits of δ, the true difference that we are seeking. In general they are set out as the estimated value $\pm t_s \times$ the S.E. of the value, where t_s is the value of t in the table for the D.F. appropriate to the S.E. and probability 100 minus the percentage confidence required.

In our example, the S.E. of the difference of the means of 10 pots is 6.05, which we worked out earlier. If we multiply this S.E. by the value of t at the 5 per cent level for 18 D.F. we get $6.05 \times 2.10 = 12.71$, and this gives us the 95 per cent confidence limits, i.e. 44.1 ± 12.71 or $31.4 \leqslant \delta \leqslant 56.8$ and can say either the true difference lies between these two values or an unlikely event has occurred, an event which has a chance of only 1 in 20 of occurring. Similarly, if we wanted to be 99 per cent confident of the value of the true mean difference, we would use the 1 per cent *t*-value. Then we have $6.05 \times 2.88 = 17.42$ and can say that we are 99 per cent confident that the true mean difference lies between $44\cdot1 - 17\cdot42 = 26\cdot68$ and $44\cdot1 + 17\cdot42 = 61\cdot52$ or $26\cdot7 \leqslant \delta \leqslant 61\cdot5$. We cannot know the mean exactly unless we measure the whole population, but this gives us much better information than using the mean of the sample alone, because we can say with a known degree of certainty within what limits it lies.

This is particularly useful if the problem is an economic one. Suppose that in the present example spraying the fungicide cost as much as 20 g of grain could be sold for. We are 99 per cent certain that the real difference is not less than 26·7, which is greater than the weight of grain required to pay for the spray, so we have at least 99 per cent confidence in saying that a user of the method will make a profit. This is a much more useful statement than saying that the effect was significant, because, although it might be significant in the statistical sense, it would be of no significance whatever in practice if it cost more to spray than the value of the extra crop produced.

There are three things to notice about confidence limits:

(1) We give names describing the certainty rather than names describing the uncertainty so, since t measures the uncertainty, we enter the t-table at the value 100 minus the percentage confidence we want, e.g. for 95 per cent confidence look up t at the $(100-95) = 5$ per cent level or for 90 per cent look up t as $(100-90) = 10$ per cent level or in tables which give the probability as a proportion, for confidence limits at $P = 0\cdot95$ look up t at $P = (1\cdot00 - 0\cdot95) = 0\cdot05$.

(2) The greater the degree of certainty the wider are the limits. In the present case, although we would be 95 per cent certain that the mean difference lay between 31·4 and 56·8, we could be only 99 per cent certain that it lay between the much wider limits of 26·7 and 61·5. Thus, although it is obviously desirable to work with a high degree of certainty, if too high a standard is set, the whole purpose may be defeated because the confidence limits will be too wide to be of any use. For example, a farmer in our example who would use the fungicide only if there was a chance of less than one in a million that he would make a loss would never use it, because no-one would ever make an experiment capable of estimating the difference to the necessary degree of certainty.

(3) There is a connection with the significance test. If the limits contain the value zero, then no difference is a reasonable estimate, which means that the null hypothesis which was that there was no difference is quite acceptable.

Experimental design

Further consideration of the t-test can help in understanding design of experiments in general. Put in algebraic terms we have,

$$t = \frac{D}{\sqrt{\frac{2\,\text{M.S.}}{r}}} = \frac{D\sqrt{r}}{\sqrt{2\,\text{M.S.}}} = \frac{D\sqrt{r}}{s\sqrt{2}}$$

where D is the difference between our treatment means, r is the number of replications of the treatments, M.S. is the error mean square, and s the standard error of a single observation. We know that the larger t is, the greater is our confidence that the null hypothesis is discredited, and the greater our confidence that the original hypothesis holds. Now t is increased by increasing D, decreasing s or increasing r. D is the difference we are measuring and is the estimate of the true difference. Obviously there is nothing that we can do to increase the true difference, but we should make sure that all other conditions are such that our estimate can be as near the true value as possible.

For instance, it was once suggested to a research student that a certain soil fungus would not grow if it was starved of nitrogen, so he compared two treatments, sterilized sand to which distilled water was added, and sterilized sand to which a solution of ammonium nitrate was added, and then inoculated both with his fungus. The fungus did not grow in either treatment, which was not surprising, since the fungus needs organic matter, phosphate, potassium and many other things, quite apart from nitrogen. So the experiment was useless—the proper treatments would have been soil containing all the fungus's requirements compared with soil containing all the requirements except nitrogen; only in this way could the effect of nitrogen be determined. This is fundamental to all biological experiments, but may be stated in terms of biological situations where the objective is concerned with growth. The effect of a treatment can be measured only if the lack of that treatment is restricting the growth.

s is the standard error of a single observation, which is derived from the deviations from the mean. We can keep this value small by removing any obvious sources of variation from the experiment. To do this we must know the material, and know in what ways it can vary, and control all variations that the hypothesis allows. In doing so care is needed not to cut down the population being investigated to such a small size that it is not worth knowing the answer. Such a situation has been known to happen with scientists testing drugs on rats who breed for themselves pure strains of in-bred rats so that the variation among them is negligible, and keep them under absolutely aseptic conditions. Their experiments have very low S.E.s, but the answers only apply to their own rats kept under their conditions. Other scientists find that they do not get the same results, and so the only use of the experiments is to promote acrimonious discussion. However, control of the error variation is an important part of experimental design and is discussed in Chapter 8.

Deciding the number of replications

r is the part of the expression which is completely under the experimenter's control, and he must always ensure that the replication is adequate for testing the sort of size difference that he expects. The number of replications affects the experimental results in two ways.

(1) It affects the number of degrees of freedom for the error term. In our case with only two treatments, if there are r replications, the error term has $2(r-1)$ D.F. In the general case of this type of experiment, if there are p treatments, the error term has $p(r-1)$ D.F., so obviously the greater r is the greater is the number of degrees of freedom for error and the more precise and reliable will this error estimate be. This precision is reflected in the table for t and indeed in all significance tests, because a lower value of t is required for any given level of significance when the number of degrees of freedom is large than when it is small (compare the 5 per cent levels, $t_{60} = 2 \cdot 00$, $t_5 = 2 \cdot 57$). Thus the difference between treatments needs to be much greater for a given level of significance and for a given error mean square if the latter is based on few D.F. than if it is based on many D.F. From this point of view it is worth noting that a change of 1 D.F. has quite a large effect on t when the D.F. are less than 10, whereas when more than 20 the effect is quite small. It is not reasonable to draw a line and say no experiment should have less than a certain number of D.F. for error, but the effect should always be borne in mind when designing experiments, and a good standard to set is to try to get the error based on at least 15 D.F. if possible.

(2) The second way replication affects the result is shown in the t-formula, and when designing experiments we make use of this fact too.

Let us design an experiment similar to the one we have just worked out. First find out how many replications are necessary to give 15 D.F. for the error. Put

$$2(r-1) \geqslant 15$$

$$r \geqslant \frac{15}{2} + 1 = 8\tfrac{1}{2}$$

so taking the nearest whole number, 9. Now see what size difference between means (D) would be significant at the 5 per cent level. Here a rough test is usually good enough, and we can take $t = 2$ and write

$$D \geqslant \frac{2s\sqrt{2}}{\sqrt{r}}$$

Now r we have just worked out as 9, but we need a likely value for s, because we shall not know its actual value until the experiment has been completed. Since s and D depend on the units of measurement used, they are rather laborious to work with. It is convenient to convert them into percentages of the mean by multiplying both sides of the expression by $100/\bar{x}$ where $\bar{x}$ is the mean for the whole experiment. Then we have

$$D\frac{100}{\bar{x}} \geqslant \frac{2s(100/\bar{x})\sqrt{2}}{\sqrt{r}}$$

Now $s \times 100/\bar{x}$ is the coefficient of variation. Denote this by C, and $D \times 100/\bar{x}$ by R, the percentage increase or response due to the treatment, and we have

$$R \geqslant \frac{2C\sqrt{2}}{\sqrt{r}}$$

and need a value for C. The great advantage of this transformation is that C, the coefficient of variation, tends to be related fairly closely to the material being experimented on, and likely values can be got from a study of the literature, e.g. the value for most plant yields is about 8 per cent, for live weight of small animals about 5 per cent, and for large animals 12 per cent.

If the literature or the experimenter's own experience does not provide an estimate, it is very easy to obtain one by taking a group of untreated material divided into about 30 units, or individuals of the same sort as will be used in the experiment, making the same measurements as will be made in the experiment and working out the sum of squares, mean square, S.E. and coefficient of variation for this group.

Suppose that we expect C to be 5 per cent. We have

$$R \geqslant \frac{2 \times 5\sqrt{2}}{\sqrt{9}} = \frac{10\sqrt{2}}{3} = \frac{14 \cdot 14}{3} \simeq 5\%$$

A difference of 5 per cent would be significant. If we decide that this is the sort of difference that makes discrediting the null hypothesis worth while, we should be prepared to carry out the experiment with 9 replications.

But suppose we really expected a smaller difference, say $2\frac{1}{2}$ per cent, we should obviously have to increase the replication greatly. Using the formula backwards we could work out how many were required. If

$$R \geqslant \frac{tC\sqrt{2}}{\sqrt{r}} \quad \text{then} \quad \sqrt{r} \geqslant \frac{tC\sqrt{2}}{R} \quad \text{or} \quad r \geqslant \left(\frac{tC\sqrt{2}}{R}\right)^2$$

with $C = 5$ per cent, $R = 2\frac{1}{2}$ per cent, and $t = 2$.

$$r \geqslant \left(\frac{2 \times 5 \times \sqrt{2} \times 2}{5}\right)^2 = (4\sqrt{2})^2 = 32$$

This emphasizes an important point in all experimentation: if small differences are to be detected, there must be considerable replication; few replications will detect only large differences. If the hypothesis is concerned with small true differences, an experiment with too few replications is worse than useless, because it confirms the null hypothesis, and therefore disputes the hypothesis even when the hypothesis is true. Since the experiment was not capable of detecting such a true difference, it is better to say that no suitable experiment was made.

This approach prevents many useless experiments being made, but it should be noted that it is not a formula which enables the experimenter to proceed with known certainty of detecting various *true* differences. Such an approach requires consideration of the variation of the coefficient of variation, and the variation of the experimentally estimated difference around the unknown true difference. Tables based on a method which takes care of the latter variation are given in Cochran and Cox (1957, pages 17-29) and will not be considered here. It suffices to say that many biologists would be amazed at the number of replications required to be even 80 per cent confident of detecting a 10 per cent difference when coefficients of variation are as high as is common in very detailed studies of parts of plants or animals, a fact which leads to many experiments failing to give any definite conclusions.

Single factor design

The experiment we have been considering is called a *one-way classification* or *single-factor design* in statistics, because the data can be classified in only one way, namely, according to the treatment, and treatment is the only factor involved. It is the simplest design and is frequently used, but should not be used if there is a better method available. It is rea appropriate to situations where little is known about the sources variation; examples are experiments with fungi or bacteria growing agar plates, experiments with enzymes breaking down particular co... pounds in test-tubes and the like.

There are two fundamental requirements when using this type of design and analysis.

(1) The experimental items must be allocated to the treatments entirely at random. Failure to do so can leave no real test of the proposed null hypothesis. For example, a young man once got the idea that elephants and similar game were responsible for the poor growth of vegetation in certain parts of Africa. He wanted to compare the growth of vegetation on land on which elephants can wander, as one treatment, with that on land on which elephants are excluded as the other. So he fenced off a single area and later marked out 10 plots inside it and 10 plots outside it. He cut and weighed the vegetation which grew in these plots, working out an error mean square from the variation between plots within the fence plus the variation between those outside the fence. Notice how this differs from our experiment in randomization. We specified that every pot must have the same chance of being sprayed with fungicide, but here the plots are not separate units for randomization of treatment. If a plot is inside the fence it must have no elephant, if outside it must have elephant. Thus the 10 plots inside are not independent of one another, so are not capable of being independent observations applicable to the treatments. When an analysis is done in the one-way classification fashion, it may be possible to say that the two areas differ, but it is not possible to say that the difference is due to keeping out elephants, because there is no test of that at all. The difference may just as likely be due to the different amounts of vegetation in the two areas to start with, or to different levels of soil fertility. The proper way to do the elephant experiment would be to select 20 areas of the bush, put a fence round each, allow game into 10 of them chosen at random, and no game into the other 10; then take cuts from each of the 20 areas to assess the growth of vegetation. Of course this would be a much more costly experiment.

This is a very common mistake, especially when dealing with expensive apparatus, e.g. comparisons of two types of machine are often made with only one representative of each type, and several replications of the test, but the error obviously includes nothing about the variation between machines of the same type, which is essential for predicting what the population of machines can do. Similarly, with controlled-environment cabinets with all the representatives of one treatment in one cabinet and all of another in a different cabinet, any differences alleged to be between treatments are equally likely to be between cabinets. In general when this design is used wrongly in this way, the differences measured between treatments tend to be larger than they should be considering the amount of natural variation measured, so hypotheses tend to be supported when they are false.

(2) The second requirement is that the data should reasonably fit this model:

$$x_{ij} = M + a_i + e_{ij}$$

where M is a value concerned with the material being used and is estimated as the mean for the whole experiment, a_i is the constant effect of the ith treatment, and e_{ij} is a variable which is normally distributed with mean zero. The distribution of t is based on this assumption so, if the assumption does not hold, then using the t-distribution to assess the probability of the null hypothesis being true will not be appropriate.

Experiments are usually too small to test rigorously how near the normal distribution the actual values of e_{ij} are, so we usually have to rely on experience. Generally, if we are dealing with a measured biological quantity with three significant figures, the assumption holds well; if we have only one significant figure, it is much more doubtful; and if values were simply integers 1 to 5, then it is a dangerous procedure to use this kind of test.

Fisher-Behrens test

One common failure of the t-test in biological situations is that the values of e_{ij} do not form a single normal distribution, but that the variation is of a different nature in the two treatments. This should not happen often in experiments in which the two samples are drawn at random from the same bulk and therefore obviously have the same variation to start with. However, there are such situations, e.g. when one of the treatments has a drastic effect on the organisms, the extreme case being a treatment which kills all the organisms in an experiment concerned with growth when the variance of that sample will be zero, whereas the other sample may have normal growth and variability. It is more common, however, when "treatments" involve different species or strains which themselves have different variation. The t-test should then be avoided, and a similar test first suggested by Behrens used instead. The statistic required is derived by considering the percentile points of two t-

distributions and depends on the D.F. of each sample and the ratio of the separate S.E.s. Tables of this statistic, called d, are given in Fisher and Yates (1963) and

$$d = \frac{\text{difference between two means}}{\sqrt{(s_1^2 + s_2^2)}}$$

when s_1 and s_2 are the S.E.s of the two respective means, is used in exactly the same way as Student's t, but the distribution of d depends on the ratio of s_1 and s_2 (put into the form of $\tan^{-1}(s_1/s_2)$ in Fisher and Yates' Tables) as well as the D.F.

Applying this to the experimental results of Chapter 3, for a_1 pots

$$\text{M.S.} = \frac{1142\cdot9}{9}$$

so, for the mean of the a_1 pots,

$$s_1 = \sqrt{\frac{1142\cdot9}{9 \times 10}} = 3\cdot56$$

and for the mean of the a_0 pots

$$s_2 = \sqrt{\frac{2155\cdot6}{9 \times 10}} = 4\cdot89$$

The ratio $s_1/s_2 = 0\cdot7281$, which to use Fisher and Yates' Tables must be converted to θ, the angle whose tangent is this quantity, when $\theta = 36°$.

$$d = \frac{44\cdot1}{\sqrt{(3\cdot56^2 + 4\cdot89^2)}} = 7\cdot29$$

Using Fisher and Yates' Table VI when $\theta = 36°$, $n_1 = 9$ and $n_2 = 9$, we should have to interpolate between $\theta = 30°$ and $45°$ and n_1, n_2 between 8 and 12. But as a first attempt we might look at the most extreme value, i.e. $\theta = 30°$, $n_1 = n_2 = 8$, when we read $d = 2\cdot294$ at the 5 per cent level and $3\cdot239$ at the 1 per cent level. Our value is much greater than this, and it is clear from the table that the value for d at $\theta = 36°$ would be less than that for $\theta = 30°$; so we would again be well satisfied that if our null hypothesis is true a very unlikely event has occurred.

Mann-Whitney test

The other common failure occurs when the data are themselves not of sufficient quality to be considered as an interval scale or when their distribution is patently not normal. Both these problems can usually be overcome by a better design of the experiment. Increasing the number of basic objects in each experimental unit will make the data approach more nearly an interval scale and, bearing in mind the Central Limit Theorem, often cause the values of e_{ij} to be more nearly normally distributed. This approach is recommended whenever possible, because biological interpretation is greatly aided when this form of analysis can be used. However, if it is not possible to overcome the problem in this way, all is not lost, since other methods, known as non-parametric methods or the use of distribution-free statistics, are available.

In the case of a single-factor design, Mann-Whitney's U would be appropriate for testing or setting confidence limits if it could be assumed that the shape of the distribution was the same for each treatment, no matter what that shape was. It is equivalent to testing if the medians of the two populations are different and putting confidence limits to the difference between the two medians, which would be very appropriate knowledge if the populations were known to have a skewed distribution. The following example will illustrate the method.

A trapping experiment was carried out in a wood to compare the density of mice in winter with the density in summer. Ten sites within the wood were chosen at random independently for each period. The null hypothesis was that the density is the same in winter as in summer.

The number of mice caught at each site is shown in Table 4.1.

Table 4.1—Number of mice caught at each of 10 sites in winter and in summer and their ranks

Winter, number of mice	31	21	15	8	13	13	11	8	18	26
Summer, number of mice	32	32	14	20	20	24	15	16	30	31
Winter, rank	$17\frac{1}{2}$	13	$7\frac{1}{2}$	$1\frac{1}{2}$	$4\frac{1}{2}$	$4\frac{1}{2}$	3	$1\frac{1}{2}$	10	15
Summer, rank	$19\frac{1}{2}$	$19\frac{1}{2}$	6	$11\frac{1}{2}$	$11\frac{1}{2}$	14	$7\frac{1}{2}$	9	16	$17\frac{1}{2}$

To make the Mann-Whitney test of this null hypothesis we must first rank the numbers caught as if they did represent a single population. Thus the smallest number is 8 and that should have the lowest rank. There are in fact two 8s, a tie for the lowest rank, so we consider that these two would occupy ranks 1 and 2, but we cannot distinguish between them so give both the rank $1\frac{1}{2}$. (In this way the sum of the ranks of n values will be the same whether there are any ties or not, which is very convenient for the arithmetic involved in the test.) Having given the 8s each $1\frac{1}{2}$, we have used up both ranks 1 and 2, so the next smallest number

(11) must have rank 3. Following this system, the two 13s share ranks 4 and 5, and so are each given $4\frac{1}{2}$, and 14 receives rank 6. The full ranking is shown in Table 4.1.

The ranks for each "treatment" are then added separately: for winter $R_1 = 78$ and for summer $R_2 = 132$. Since the equivalent of the first 20 integers has been used, the total of the ranks for the two treatments can be checked using the well-known formula for the sum of the first n integers $= \frac{1}{2}n(n+1)$. In this case $n = 20$ so the sum $= \frac{1}{2} \times 20 \times 21 = 210$, and $R_1 + R_2 = 78 + 132 = 210$, so all is well.

Then Mann-Whitney's U is worked out for both treatments using the formulae

$$U_1 = n_1 n_2 + \tfrac{1}{2} n_2 (n_2 + 1) - R_2$$

and

$$U_2 = n_1 n_2 + \tfrac{1}{2} n_1 (n_1 + 1) - R_1$$

where n_1 is the number of observations in the first sample (i.e. the sample whose ranks sum to R_1) and n_2 is the number in the second sample (sum of ranks, R_2). It should be noted that the test does not require the number of observations to be the same for both samples, though in an experiment it is usual that they would be.

In the present example

$$U_1 = 10 \times 10 + \tfrac{1}{2} \times 10 \times 11 - 132 = 23$$

and

$$U_2 = 10 \times 10 + \tfrac{1}{2} \times 10 \times 11 - 78 = 77$$

Again there is a check; by adding the two algebraic expressions for U_1 and U_2 and substituting $\frac{1}{2}(n_1 + n_2)(n_1 + n_2 + 1)$ for $R_1 + R_2$ for the reason explained when checking the calculation of the Rs, we find that $U_1 + U_2 = n_1 n_2$. Thus in the present example $U_1 + U_2$ should be $10 \times 10 = 100$ and indeed $23 + 77 = 100$, so the arithmetic checks.

The smaller U is now referred to Mann-Whitney's tables (Siegel, 1956, pp. 274–77) which give the smallest value of the smaller U which can arise due to chance at various levels of probability if the null hypothesis is true. (When the replication is the same for both samples, the smaller U comes from the formula with the larger R, so only one U need be calculated; but with unequal replication it is not so obvious which is the smaller U, so both must be calculated.) In this case if the null hypothesis were that the median number of mice does not differ from winter to summer, we must use the tables for two-tail tests. We find the critical Us for $P = 0.002, 0.02, 0.05$ and 0.10 when $n_1 = 10$ and $n_2 = 10$ to be 10, 19, 23 and 27 respectively. Our smaller U is 23 so, if the number of mice is the same in winter and summer, we have witnessed an event which can occur in only 5 per cent of such tests. Thus we would most likely prefer to believe that the number is not the same in the two seasons.

Like the statistic t, U can also be used for setting confidence limits. In the present case it might be useful to be able to say with some degree of certainty what the difference between the median number caught in winter and in summer is likely to be. To do this we must first set out the sample values in ascending order in a two-way array and obtain all the differences between individual values. Thus, subtracting winter from summer in each case, we obtain the values in Table 4.2.

Table 4.2—All possible differences in numbers of mice caught between samples taken in summer and those taken in winter

Number caught in summer samples

	14	15	16	20	20	24	30	31	32	32
8	+6	+7	+8	+12	+12	+16	+22	+23	+24	+24
8	+6	+7	+8	+12	+12	+16	+22	+23	+24	+24
11	+3	+4	+5	+9	+9	+13	+19	+20	+21	+21
13	+1	+2	+3	+7	+7	+11	+17	+18	+19	+19
13	+1	+2	+3	+7	+7	+11	+17	+18	+19	+19
15	−1	0	+1	+5	+5	+9	+15	+16	+17	+17
18	−4	−3	−2	+2	+2	+6	+12	+13	+14	+14
21	−7	−6	−5	−1	−1	+3	+9	+10	+11	+11
26	−12	−11	−10	−6	−6	−2	+4	+5	+6	+6
31	−17	−16	−15	−11	−11	−7	−1	0	+1	+1

(Number caught in winter samples)

Then supposing we wished to be 95 per cent confident of our statement we look up U for $P = 0.05$ (i.e. one minus the confidence required, $1 - 0.95$ in this case) with n_1 and n_2 the number of observations in the two samples. We have seen that $U = 23$ for $n_1 = 10, n_2 = 10$ and $P = 0.05$ and, using this, we find the 23rd highest value and 23rd lowest value in the array. Thus, starting at the highest value $(+24)$ in the top right-hand corner, we can quickly see that the 20 highest values occur in the last four columns of the first five rows. There are then two $+17$s in the sixth row to make the 21st and 22nd, and so the 23rd value must be one of the $+16$s. Likewise to find the 23rd lowest value we see straightaway that there are 22 negative values, so one of the zeros must be the 23rd. We can then say with 95 per cent confidence that $0 \leqslant$ difference $\leqslant 16$ or, put another way, that either the true difference between the medians lies between 0 and 16 or an unlikely event has occurred.

The example further illustrates the relationship between tests and confidence limits. The probability level obtained for disputing the null hypothesis was exactly $P = 0.05$ and, by using 95 per cent confidence limits, zero difference was just included. Thus the latter could equally well be used as a test, though with the Mann-Whitney procedure it is much more tedious to calculate.

CHAPTER FIVE

EXPERIMENTS WITH MORE THAN TWO TREATMENTS

In the previous chapter we considered an experiment in which there was but one null hypothesis which could be tested by seeing how reasonable it is to consider the difference between the particular measures of two populations to be zero, given the data obtained from samples from those populations. In serious biological research, however, it is often necessary to consider more than one null hypothesis at the same time, and this involves more than two treatments in an experiment. At the outset it should be stressed that the approach made still requires that each question is put in a precise form. The attitude of including many treatments in an experiment in the hope that something will become established is not very fruitful, and the usual statistical methods are not appropriate, because the situation necessarily degrades into one of saying, "This looks a likely difference, let's test it". This really tests only the experimenter's ability to make statistical calculations roughly in his head, since it is only because he can estimate that a difference is significant that he tests it formally to see if it is.

However, there are many occasions requiring more than two treatments when precise questions are asked and can be answered rigorously. It is often necessary to ask the question, "Can all these supposedly different populations be considered to be all the same population?" an important question in genetics and taxonomy. As an example, suppose that a botanist observes that the stalks of primroses are of different lengths in different habitats. He immediately sees two possibilities: the environment may affect length of stalk or special genotypes may have developed in each area. To distinguish between these interpretations, he could grow samples of each, all in the same environment, setting his null hypothesis that all these primroses are from the same population. Let us assume he has three habitats, h_0, h_1 and h_2, and that he can draw a random sample of plants from each, and grow them in good potting loam in pots in a glasshouse under uniform moisture and other environmental conditions. He must first decide how many replications or plants per sample he should use. As explained earlier, there should be sufficient degrees of freedom for

EXPERIMENTS WITH MORE THAN TWO TREATMENTS

the error term, which must obviously come from variation between plants from the same habitat. Using the one-way classification design

$$\text{error D.F.} = p(r-1)$$

when p is the number of treatments and r the number of replicates as shown on p. 44. Thus with three habitats as treatments, to achieve at least 15 D.F. for error

$$3(r-1) \geqslant 15 \quad \text{or} \quad r \geqslant 6$$

It is unlikely that the experimenter would have, or could find in the literature, any reasonable estimate of the coefficient of variation of primrose stalks to enable him to forecast what sort of differences his experiment could detect. He could, of course, make a preliminary experiment to find out; but in a case like this, where the preliminary experiment would involve about the same amount of work as the main experiment, it would be sensible to proceed with the experiment with the six replications, bearing in mind that, if the coefficient of variation turns out to be very large and the null hypothesis is not disputed, it is better to forget the experiment and say that no suitable test has been made, rather than that there is evidence in support of the null hypothesis. A further, more precise experiment, with sufficient replication to detect the sort of differences that are thought to be important, should then be designed using this knowledge of the coefficient of variation.

Suppose then that there were six replications in a one-way classification design and that the mean lengths of stalk are those given in Table 5.1.

Table 5.1—Mean length of primrose stalks from three habitats

Habitats	Length (mm)						Total (T)	Mean
h_0	83	82	98	76	66	64	469	78·2
h_1	106	96	107	94	87	92	582	97·0
h_2	81	93	79	98	111	76	538	89·7
							1589	88·3

When there were two treatments, we tested if there was a difference between the population means by comparing the differences between the two sample means with the natural variation which was estimated from the variation between members having the same treatment. With more than two treatments we cannot get a single difference to represent the treatment effect, but, remembering the argument about assessing natural variation as the square of the deviations of individuals from the mean of a treatment, it is not unreasonable to think that we could assess the variation between treatments as the square of the deviations of the treatment means from the general mean. Thus, if there was no natural variation, and if treatments had no effect, we might expect h_0, h_1 and h_2 all to have a mean of 88·3 in this case, and any deviation from that value would be a measure of the treatment effect. In fact it is usually easier to work with totals than with means, but following this argument we require two estimates of variation, one of which will include the treatment effects and one which is just natural variation. We could calculate the latter in the same way as was used in the two-treatment case, i.e. calculate the S.S. for each treatment separately, add the three lots of S.S., and obtain the mean square by dividing by the D.F. ($3 \times 5 = 15$, in this case); but where there are more than two treatments it is usual to find the error mean square by computing what is called an *analysis of variance*.

Analysis of variance

An analysis of variance consists of writing down a table like Table 5.2 (p. 58) in which the first column specifies the sources of variation in the experiment. In this case there are two which can be specified and make up the total variation: (1) variation due to habitats and (2) natural variation within habitats which we shall call *error*, and these make up the total variation in the experiment. The second column is headed D.F. and we write here the degrees of freedom for each source of variation. Since there are three habitats, the variation between them will have $3 - 1 = 2$ D.F. as we saw on pp. 27 and 32. The error variation has 5 D.F. for the variation within each of three habitats, so 15 D.F. altogether. The D.F. should now be added ($2 + 15 = 17$) and checked that it is the same as the total D.F. (3 treatments × 6 pots per treatment = 18 pots; hence 17 D.F. for total variation). It is obvious that, since all the variation is represented by the total, it is possible to obtain the D.F. for one of the items by difference between the total and all the others, but it is not advisable to do so, since there would then be no check that all sources of variation had been accounted for.

In the third column we write the sum of squares for each source of variation. Taking the habitats first, we have the totals for h_0, h_1 and h_2 and, calling these T_1, T_2, T_3, the sum of squares of deviations of these totals from their mean would be

$$\sum T^2 - \tfrac{1}{3}(\sum T)^2$$

but this would be the sum of squares of totals of 6 pots and, to make our sums of squares column additive, all items need to be in the same units. Since the total sum of squares will come from deviations of single pots, it is best to bring all sums of squares to that level. Remembering that the variance of a sum of r values is r times the variance of a single value (p. 31), when estimating these variances we can calculate $\text{M.S.}_{\Sigma x} = r\text{M.S.}_{.x}$. Since mean squares are simply sums of squares divided by degrees of freedom, it follows that

$$\text{S.S.}_{\Sigma x} = r\text{S.S.}_{.x} \quad \text{or} \quad \text{S.S.}_{.x} = \frac{\text{S.S.}_{\Sigma x}}{r}$$

To obtain our sum of squares for treatments in units of single pots, we must divide the sum of squares of the deviations of the treatment totals by the number of pots used to make each total. In the present case

$$\text{S.S. for habitats} = \frac{\Sigma T^2 - \frac{1}{3}(\Sigma T)^2}{6}$$

$$= \tfrac{1}{6}\Sigma T^2 - \tfrac{1}{18}(\Sigma T)^2$$

It will be noticed that the calculation can be done in two parts: $\tfrac{1}{6}\Sigma T^2$ is the sum of the squares of the totals divided by the number of individual objects which make up each total; and $\tfrac{1}{18}(\Sigma T)^2$ is the usual correction factor which will be used when calculating the total sum of squares, since the sum of the treatment totals must equal the grand total, and the divisor ($18 = 3 \times 6$) is the number of treatments times the number of pots in each treatment, i.e. the total number of observations.

Thus, in general, when all treatments are replicated to the same extent, allowing there to be r replications of each,

$$\text{S.S. for treatments} = \frac{\Sigma T^2}{r} - \frac{(\Sigma x)^2}{n}$$

If the replication is not the same for all treatments, the same principle holds but the calculation becomes

$$\frac{T_1^2}{r_1} + \frac{T_2^2}{r_2} + \ldots - \frac{(\Sigma x)^2}{n}$$

where r_1 is the number of replications of the treatment whose total is T_1, r_2 the number for T_2, and so on.

In the present case the S.S. for habitats

$$= \frac{469^2 + 582^2 + 538^2}{6} - \frac{1589^2}{18}$$

$$= \frac{848\,129}{6} - \frac{2\,524\,921}{18}$$

$$= 141\,355 - 140\,273 = 1082$$

The total sum of squares is obtained as

$$\sum x^2 - \frac{(\sum x)^2}{n}$$

i.e.

$$83^2 + 82^2 + \ldots + 76^2 - 140\,273 = 143\,367 - 140\,273 = 3094$$

Both these sums of squares can be entered in the analysis of variance table, and the error S.S. obtained by difference, i.e.

$$3094 - 1082 = 2012$$

The next column should contain the mean squares necessary for making the test. In this case we wish to compare the variation between the habitats with the natural variation which we have estimated as the variation between pots of primroses from the same habitat, i.e. the variation given in the error line of the table. Thus we want the M.S. for habitats = S.S./D.F. = 1082/2 = 541 and the M.S. for error = 2012/15 = 134. It would be a waste of time to calculate the M.S. for total, since this represents the mixture of variation which we have tried to separate into parts with real biological meaning.

Table 5.2—Analysis of variance

Sources of variation	D.F.	S.S.	M.S.	F
Habitats	2	1082	541	4·04
Error	15	2012	134	
Total	17	3094		

Now the test to be used is called the *variance ratio* or *F*-test, where

$$F = \frac{\text{mean square of the item to be tested}}{\text{mean square of the error}}$$

In the present case

$$F = \frac{\text{M.S. of habitats}}{\text{M.S. of error}} = \frac{541}{134} = 4 \cdot 04$$

Like t-tables, tables of F have been worked out to give the probability of normally distributed random variates with the same variance for each group leading to Fs of various sizes. Detailed tables are given in Fisher and Yates (1963) and a small range of values in Appendix Table II (p. 226). The tables have the same logic as t-tables in that the values given are the largest which can arise at the level of probability stated. If we are prepared to assume that our data can be considered to be normally distributed random variates, with the same variance for each habitat, then, if our experiment gives a larger value, we can say that, if there is not a true difference between the variation between treatments and the natural variation, we have witnessed an event which can occur less frequently than the chosen level. Whereas the t-tables are two-dimensional, in the sense that each value depends upon the number of degrees of freedom for the error and the level of probability, the F-tables are three-dimensional, and each value depends upon the degrees of freedom of treatments, as well as the degrees of freedom for error and the level of probability. In the present case we enter the table in the column $n_1 = 2$, the number of degrees of freedom for our treatments and the row $n_2 = 15$, the degrees of freedom for error and read $F = 3 \cdot 68$ for $P \leqslant 0 \cdot 05$ and $6 \cdot 36$ for $P \leqslant 0 \cdot 01$. The F-value from the experiment is $4 \cdot 04$, so we can conclude that if the null hypothesis is true, and these three habitats represent a single homogeneous population of primroses, we have witnessed an event which has a probability of occurring of less than 5 per cent. We would prefer to believe that it is not true and that they do not form a homogeneous population. The experimenter would therefore be prepared to continue his investigation on the assumption of genetic effects rather than only environmental effects. He would be wise to work out the coefficient of variation to see what sort of difference the experiment was capable of detecting, as this would aid him in designing other experiments:

$$\text{C.V.} = \frac{\sqrt{(\text{error M.S.})} \times 100}{\text{general mean}} = \frac{\sqrt{(134)} \times 100}{88 \cdot 3} = 13\%$$

This is not a particularly high value for a small part of a plant, but it shows that considerable replication would be required to detect small differences.

The relation between F and t

It will be seen that there is great similarity between an F-test and a t-test, and indeed it would be silly if they gave different answers in situations where either could be used. A good explanation of the relation between the four common tests based on the assumption of normally distributed data is given by Mather (1964, pp. 46–49). It will suffice here to see the connection between the test in the primrose example and that of the wheat experiment in Chapter 4. If we had used the F-test in the wheat example, we should first calculate the sum of squares for treatments. Here there were only two treatments, and it was shown on p. 25 that, when there are only two values in a sample, the S.S. = $\frac{1}{2}$ (difference between them)^2. Working on treatment totals it will again be necessary to divide by the number of individuals that have been added to make these totals, so, calling the difference between the two treatment totals D and the number of replications of each total r,

$$\text{S.S. for treatments on the basis of single observations} = \frac{D^2}{2r}$$

and since treatments have only 1 D.F., the M.S. for treatments will be

$$\frac{D^2}{2r}$$

The error mean square will be the same as before and, calling it M.S.$_E$,

$$F = \frac{D^2}{2r \, \text{M.S.}_E}$$

We calculated that

$$t = \frac{\text{difference}}{\text{S.E. of the difference}} = \frac{D}{\sqrt{(2r \, \text{M.S.}_E)}}$$

so

$$t^2 = \frac{D^2}{2r \, \text{M.S.}_E} = F$$

Comparisons in Appendix Table II (p. 226) will show that when $n_1 = 1$, $F = t^2$ for all values of n_2 and for all probabilities, so the answer will be exactly the same for two treatments whichever test is used; which to use depends on the problem. The F-test is simpler if all that is required is a statement of the experimenter's belief in the null hypothesis, but more usually he would wish to display means and standard errors, and talk about the size of the difference; then having to calculate the standard error, there is no advantage in performing the extra operations required for the F-test.

Like the t-test, the F-test is based on certain assumptions which include the requirement that the data should follow a normal distribution, and that the variance should be the same for each treatment. If the data themselves do not comply with the latter requirement, there are often ways of transforming them so that the transformed values do.

Kruskal-Wallis procedure

If the data are such that they cannot be made to fit the assumptions for F, a non-parametric test of the same sort exists and is known as the Kruskal-Wallis procedure. Though not as powerful as the F-test, it simply requires the assumption that the distributions of the different treatments are all of the same shape, the same assumptions as for Mann-Whitney's U-test for two treatments. Like Mann-Whitney's test it assesses if the medians of the population are the same. The following example will show the procedure.

A soil microbiologist wished to know whether or not populations of wild white clover differ in degree of nodulation between three different ecological sites. He took random samples from each site and assessed the number of nodules per plant in each instance. His results are shown in Table 5.3.

Table 5.3—Number of nodules per clover plant and ranks for three sites

Site A Number	Rank	Site B Number	Rank	Site C Number	Rank
43	14	15	4	18	7
56	15	10	2	26	9½
30	11	12	3	31	12
17	5½	17	5½	7	1
26	9½	24	8	36	13
Total ranks (R)	55		22½		42½

The observations are first ranked as if they formed one large sample, and these ranks are totalled for each sample. Denoting these totals by $R_1, R_2, \ldots$, and letting the number of observations in each sample be $n_1, n_2, \ldots$, we have a check on the ranking because $\sum R_i = \frac{1}{2}\sum n_i(\sum n_i + 1)$. The test statistic, denoted by H, is calculated as

$$H = \frac{12}{\sum n_i(\sum n_i + 1)} \sum \frac{R_i^2}{n_i} - 3(\sum n_i + 1)$$

in the present case,

$$H = \frac{12}{15 \times 16}\left(\frac{55^2}{5} + \frac{22\frac{1}{2}^2}{5} + \frac{42\frac{1}{2}^2}{5}\right) - 3 \times 16 = 5\cdot 38$$

Siegel (1956, pp. 282–3) gives a table showing the probability of obtaining various values of H when the population medians are equal for three treatments with ns from 2, 1 and 1 to 5, 5 and 5. In the present case with 5 observations in each treatment the table gives

H	P
8·00	0·009
7·98	0·010
5·78	0·049
5·66	0·051
4·56	0·100
4·50	0·102

Our H is 5·38 with a probability less than 0·10 but greater than 0·05, so the evidence against the null hypothesis that site does not affect nodulation would be considered quite strong, and the microbiologist would be well advised to continue with his study.

If the treatments have more than five observations H approximates to χ^2 with $(p-1)$ D.F. (Appendix Table III, p. 228) where p is the number of treatments.

Tests within a set of treatments

Although the t-test was published first (Student, 1908), the F-test (Snedecor, 1934) based on a similar z-test by Fisher (1925) is of more general application in theoretical statistics and for a long time was used extensively by biologists, who considered most of their experiments to be in the form of the primrose experiment that we worked out. However, the type of question asked in that example does not often end the investigation. Knowing that genetic differences exist, the botanist would probably like to proceed and group the habitats of particular genotypes, and find out what type of environment is conducive to causing long-stalked primroses to arise, or determine if there are genotype × environment interactions. Just knowing that the populations differ will not get him very far. To take another example, the biochemist interested in quantity of coumestrol in clover will not be satisfied with simply knowing that the various cultivars of clover differ in coumestrol concentration but will wish to know *which* consistently give levels high enough to affect the animals that eat it.

To do this we require tests for making individual comparisons within a set of treatments. A first thought might be to test all possible differences with a t-test. Bearing in mind that the t-test is based on a definite null hypothesis, it is not surprising that the theory breaks down if it is applied haphazardly without such a basis. There are two difficulties.

(1) If the tests are not independent of each other, then the result of one is affected by the result of another, e.g. if we compare A with B and A with C, and A is small by chance, then the claim that B and C are greater than A are not two separate pieces of information—because, if B is greater than A here, C has a greater chance of being assessed as greater than A than it would have in a separate experiment.

(2) There will always be tests that give too much weight to the chance variation, e.g. there will be one which tests the highest against the lowest, and these are likely to contain positive and negative deviations, respectively. In fact, it can be shown that if there is no true difference between three populations, the difference between the highest and lowest means of samples will appear significant at the 5 per cent level in about 13 per cent of experiments.

Statisticians have devised a number of improvements in recent years, but it is the biologist who can best get over the problem by sensible experimental design. It is difficult to think of a biological situation in which we cannot do better than compare a number of treatments one with another. What does one learn from such a test? Since we have to draw a line of belief based on the unlikeliness of differences if the populations are the same, we are faced with problems such as: we believe that A is greater than C, and that A is the same as B, and B is the same as C. Statistics has nothing to offer as a solution to that piece of logic.

Statisticians have provided a number of tests called *multiple range tests* for *a posteriori* testing which allows for some of the probability inadequacies of t-tests, but these should be used by a biologist only in very rare circumstances when no *a priori* hypotheses are possible. In practice this means at the very early stage of an investigation when he is searching for treatments; as soon as some suggestion of which treatments are likely to solve the problem is obtained, he should design a new experiment with appropriate *a priori* hypotheses.

Duncan's multiple range test

The following example of the use of Duncan's multiple range test (Duncan, 1955) shows the calculations involved in *a posteriori* testing. In a physiological study it may be necessary to assess different methods of exciting a muscle so that we have a method suitable for comparing the responses of different muscles from the same animal or the same muscle from different species. The first experiment might involve five quite different methods of stimulation with, say, five replications of each method.

Suppose the experiment gave the following treatment mean responses measured in suitable units:

Methods	A	B	C	D	E
Mean response	31·3	33·1	35·4	38·3	37·1

An analysis of variance would be carried out as on p. 58. This time there are five treatments, so 4 D.F. for treatments and five replicates of each. Then $5 \times 4 = 20$ D.F. for error making a total of 24 D.F., one less than the total number of observations. The S.S. column must be completed as before, the devisor for $\sum T^2$ will be 5 this time, but the only mean square required is that for error, which turns out to be 11·40. To use Duncan's multiple range test we must first calculate the difference between all pairs of treatments. This is best done by setting up a table like Table 5.4.

Table 5.4—Duncan's multiple range test

	D	E	C	B	Mean	Treatment
	38·3	37·1	35·4	33·1		
$D_5 = 4·91$	7·0*	5·8*	4·1	1·8	31·3	A
$D_4 = 4·80$	5·2*	4·0	2·3		33·1	B
$D_3 = 4·68$	2·9	1·7			35·4	C
$D_2 = 4·45$	1·2				37·1	E

Along the top margin the treatment means are arranged in descending order, omitting the lowest, and down the right-hand margin they are arranged in ascending order, omitting the highest.

Now we subtract the row value from the column value for all the cells in the table which are not below the main diagonal. By doing this we produce all the differences between individual treatments. Thus starting in the first row of column 1 we have Treatment D − Treatment A = 38·3 − 31·3 = 7·0; in the second row of column 1, Treatment D − Treatment B = 38·3 − 33·1 = 5·2 and so on giving the complete set as shown.

We then require the S.E. of the treatment means. Since each treatment had five replications this is

$$\sqrt{\frac{\text{M.S. for error}}{5}} = \sqrt{\frac{11\cdot40}{5}} = 1\cdot51$$

Duncan's tables, part of which are given in Appendix Table IV (p. 228), are three-dimensional in that each value is defined by the number of degrees of freedom for the M.S. from which the S.E. was obtained (n_2), the distance apart in the range of the means being compared ($p = 2$ meaning that the means are adjacent, $p = 3$ that there is one in between, and so on) and the level of probability. In the present example, using the 5 per cent level of significance, we look along the row with $n_2 = 20$ and multiply the values in the columns $p = 2, 3, 4$ and 5 by our S.E. in turn to produce least significant differences (D) for the four possible range relationships among five values. Thus

$$D_2 = 2\cdot95 \times 1\cdot51 = 4\cdot45$$
$$D_3 = 3\cdot10 \times 1\cdot51 = 4\cdot68$$
$$D_4 = 3\cdot18 \times 1\cdot51 = 4\cdot80$$
$$D_5 = 3\cdot25 \times 1\cdot51 = 4\cdot91$$

These are then written in order, starting at the bottom of the left-hand margin.

Now each D-value applies to all the differences on the diagonal next to it. Thus D_2 applies to the differences between D and E, E and C, C and B, B and A, all of which are adjacent values in the range. D_5 on the other hand applies only to the difference between D and A which are the highest and lowest values respectively. The test now is very simple: we just see which differences are greater than their appropriate D-value. We start with D_5 and put a star over the first value to show it is greater than D_5; then work through D_4 and so on. We do it in this order because, when we get to a value which is not significant, we know that there can be nothing significant in the cells which form the square below it and to the right of it.

If we wanted a more severe test we could repeat the process using Duncan's 1 per cent probability values. However, using the 5 per cent values we will now sum up our findings. It is usual to set out a table of treatment means in descending order, and to include the S.E. of the treatment means:

Treatment	D	E	C	B	A	S.E.
Mean	38·3	37·1	35·4	33·1	31·3	1·51

Either at the end of the table or below it, make a statement about the treatments considered significantly different, either by writing D, E > A and D > B or by writing

$$\underline{D\ E\ C\ B\ A}$$

by which we mean that treatments joined by any line do not differ significantly. Thus D, E, C are not significantly different from one another, nor are E, C, B nor C, B, A. We are left with the conclusion that any treatments not joined by a line are significantly different.

This method is similar to using numerous t-tests, except that the values used are nearer the proper probability values than they would be if we used Student's table and the same value for every comparison. It is interesting to note that the value under $p = 2$ is always $t\sqrt{2}$ for all D.F. Since in Duncan's test we have least significant difference $= D\sqrt{(M.S./r)}$, whereas in the t-test, L.S.D. $= t\sqrt{(2\,M.S./r)}$, the two tests are exactly the same for values adjacent in the range. However, the method does not always help the experimenter who is asking the experiment which treatment it is worth continuing to study. If he finds that one is outstandingly greater than the rest, e.g. $A > B, C, D, E$, he may be satisfied because he can discard the rest and concentrate future work on that one. If he has a situation like ours, he can conclude that D gives a high value and A a low value; but what is he to say about C which does not differ significantly from either of the two extremes? Thus he is left with little more information than he had before he did the test. For this reason many would say that this sort of test is suitable only for preliminary investigation. Armed with this information, the biologist should now use his biological thinking powers and produce unequivocable hypotheses which can be tested rigorously before pronouncing on the subject to the rest of the world.

Dunnett's test

Even in the earliest stage of an investigation it is usually possible to do something better. We can use an approach containing some *a priori* hypotheses which enable the experimenter to nominate the tests before the experiment starts. There should be no more tests than there are degrees of freedom for treatments. This is similar to considering that we have several experiments combined into one, but are economizing in work by obtaining a single estimate of the natural variation. In the early stages of an investigation it is often possible to include a "nil" treatment or "control"

EXPERIMENTS WITH MORE THAN TWO TREATMENTS

treatment. For example, a treatment which allows the organism to grow naturally may be included, and the various experimental treatments can then be compared one at a time with natural growth; this is a simple extension of the experiment designed in Chapter 2. Again, in the applied field, there will often be some well-known and well-tried practice which can act as a "control", and with which suggested new practices can be compared.

There is still the problem of lack of independence between the comparisons, but Dunnett (1955) has produced tables of t which enable us to declare our confidence in the joint statement that we make when comparing several experimental treatments with the same control. To illustrate the method, suppose that in the previous experiment (p. 64), treatment A was a well-known method of stimulating muscle, and B, C, D, E were new methods thought up by the experimenter. He would wish to know which of these gave responses different from the well-known method. In statistical parlance he would wish to test each against the control, setting as null hypotheses A = B, A = C, A = D and A = E. The test is made by using a least significant difference based on the appropriate t-value in Dunnett's table, a small range of which is shown in Appendix Table V (p. 229). These tables are again three-dimensional. Each value is defined by the number of degrees of freedom (n_2) of the mean square from which the S.E. of the difference between the two means is derived, the number of experimental treatments excluding the control (p) which is the same as the number of tests to be made, and the level of confidence required. The L.S.D. is obtained in the same way as with Student's t by simply multiplying the S.E. of the difference by the appropriate t taken from Dunnett's table. In the present case, using the 5 per cent level of probability and entering Appendix Table V at $n_2 = 20$ and $p = 4$, t is 2·70 so

$$\text{L.S.D.} = 2 \cdot 70 \times \sqrt{\frac{2 \text{ M.S.}}{5}} = 5 \cdot 77$$

We can set out the differences between the means thus:

$$B - A = 1 \cdot 8$$
$$C - A = 4 \cdot 1$$
$$D - A = 7 \cdot 0 *$$
$$E - A = 5 \cdot 8 *$$
$$\text{L.S.D.} (P = 0 \cdot 05) = 5 \cdot 77$$

placing an asterisk against the values which we consider significantly different from zero. We have 95 per cent confidence in the whole statement that methods D and E give greater responses than the standard method A, but methods B and C do not. The physiologist then has something definite to work on for his next experiment, e.g. he might be prepared to use either of methods D or E, depending on which was simpler, or might design a new experiment to compare these two. This sort of screening process is very useful in many biological situations. The relation with Student's t should be noticed; when $p = 1$, i.e. when there are only two treatments, the control and one experimental treatment, Dunnett's t = Student's t, so it is the same test. When there is more than one experimental treatment, Dunnett's t is larger than Student's t. Indeed if we had used Student's t the L.S.D. would have been only 4·45 but, using that value, we should not know the exact level of confidence that we could place on all our statements. It would certainly be less than 95 per cent.

The Dunnett tables can be used for setting confidence limits in the same way as Student's, and in the present case we could say with 95 per cent confidence that the differences from the standard method for the various new ones are:

    B   1·8 ±5·77, i.e. between −3·97 and +7·57
    C   4·1 ±5·77, i.e. between −1·67 and +9·87
    D   7·0 ±5·77, i.e. between +1·23 and +12·77
    E   5·8 ±5·77, i.e. between +0·03 and +11·57

Notice that we set the null hypothesis in the form: B does not differ from A, and this, which is called a two-sided test, is commonly the form required in fundamental research. In the screening process, especially in the commercial situation, the object is often to find something better than is present already, and anything worse than the control is not expected. One-sided tests enable us to set null hypotheses of the form: B is not greater than A, C ≯ A and so on. The second half of Appendix Table V (p. 230) gives the Dunnett t-values for one-sided tests. The 95 per cent value for $n_2 = 20$, $p = 4$ is 2·30, so

$$\text{L.S.D.} = \sqrt{\frac{2 \times 11\cdot 40}{5}} \times 2\cdot 30 = 4\cdot 91$$

a smaller quantity, which is not surprising since the statistic is derived from the extremes of only one side of the distribution. When testing, only values greater than +4·91 would be declared significant; values less than −4·91 would not be remarked upon as differing from zero as they would be in the two-sided test. When setting confidence limits, the confidence interval would be only subtracted from the mean, not both added and subtracted as in the two-sided case. In the present example we could say with 95 per cent confidence that

B exceeds A by at least $1·8 - 4·91 = -3·11$
C exceeds A by at least $4·1 - 4·91 = -0·81$
D exceeds A by at least $7·0 - 4·91 = +2·09$
E exceeds A by at least $5·8 - 4·91 = +0·89$

Similarly one-sided tests can be used for hypotheses in the form: B is not less than A, when only values less than minus the value in the table would be significant, and the confidence limit would be formed by adding the L.S.D. to the difference $B - A$. Where Student's t is appropriate it can also be used for one-sided tests; then, using tables such as Appendix Table II, the probability levels are halved, i.e. the 5 per cent values indicate significance at $P = 0·025$, etc. It should, however, be stressed that which test is used must be decided by the hypotheses which are set before the experiment starts, not because of some peculiarities seen in the data.

Replicating the control treatment

When several experimental treatments are compared with a single control treatment, it is desirable to ensure that the control value is well based. Often it is worth including more replications of the control than of each of the experimental treatments. The control is going to be used in all comparisons and, therefore, if its value is a long way from the true value of the population, all the tests are going to give poor information about the new treatments. Replicating the control treatment more than the others can give lower S.E.s for the tests, for a given amount of work, than having all treatments replicated to the same extent.

Consider an experiment in which there are p experimental treatments, each replicated r times, and a control treatment replicated ar times. Suppose there are C units available for the experiment. If we use them all,

$$C = ar + pr$$

and the S.E. of the difference between the mean of an experimental treatment and the mean of the control treatment will be

$$\sqrt{\text{M.S.}_E \left(\frac{1}{ar} + \frac{1}{r}\right)} = \sqrt{\frac{\text{M.S.}_E}{r}\left(\frac{1}{a} + 1\right)}$$

But using a total of C units,

$$r = \frac{C}{a+p}$$

so

S.E. of the difference $= \sqrt{\dfrac{\text{M.S.}_E}{C}\left(\dfrac{1}{a}+1\right)(a+p)} = \sqrt{\dfrac{\text{M.S.}_E}{C}\left(1+\dfrac{p}{a}+a+p\right)}$

For a particular experiment M.S._E, C and p are constants, and this S.E. will be a minimum when $[1+(p/a)+a+p]$ is minimum. Differentiating this expression with respect to a and setting the derivative equal to zero, we have

$$-\frac{p}{a^2}+1=0 \quad \text{or} \quad a=\sqrt{p}$$

In other words, we get the lowest S.E. and consequently can detect the smallest difference between each of p experimental treatments and a control treatment, for a given amount of material, if we replicate the control treatment $\sqrt{p}$ times the number of replications of the other treatments. For example, if we had four new treatments and a control, we would make better use of 30 units by replicating the four new treatments five times, and the control 10 times, than by replicating all of them six times. When this is done the analysis of variance needs care, but is not much more difficult. To obtain the S.S. for treatments, we must use the more general formula given on p. 57 and in this case if T_1 were the total for the control (a total of 10 individuals) and T_2, T_3, T_4 and T_5 totals of five individuals each for the other treatments.

$$\text{Treatment S.S.} = \frac{T_1^2}{10} + \frac{T_2^2+T_3^2+T_4^2+T_5^2}{5} - \frac{(\sum x)^2}{30}$$

The error S.S. can be obtained as the difference between the total S.S. and treatment S.S. in the usual way.

It is often necessary to use a control treatment in a different way. Suppose we wish to test two or more chemicals supplying the same nutrient. With nitrogen as the nutrient, we might want to see if there was a different response to applying ammonium salts from applying nitrates. If we did a simple experiment with two treatments and found no significant difference, this might be because they were equally effective, or it might be because there was so much nitrogen already present in the medium that neither had had any effect at all. Then the conclusion that they had the same effect would be useless as a prediction for the future. If we had three treatments, the other being a control of no nitrogen applied, we should then know which of the two possible effects we had measured, because we could see if there was a response to nitrogen as well as if the two substances differed in effect. So in this situation we should need to compare the mean of all the experimental treatments with the control to see if on average the application of nitrogen had an effect.

In this case using the same symbols and assumptions as before the S.E. that we wish to minimize is

$$\sqrt{\text{M.S.}_E \left(\frac{1}{ar} + \frac{1}{pr}\right)}$$

which on substituting for r becomes

$$\sqrt{\frac{\text{M.S.}_E}{C}\left(\frac{1}{a}+\frac{1}{p}\right)(a+p)} = \sqrt{\frac{\text{M.S.}_E}{C}\left(2 + \frac{a}{p} + \frac{p}{a}\right)}$$

Equating the derivative with respect to a of $[2+(a/p)+(p/a)]$ to zero gives

$$\frac{1}{p} - \frac{p}{a^2} = 0 \quad \text{or} \quad a = p$$

i.e. the most efficient use of our material as far as this test is concerned is to divide the material equally between the control on the one hand and the experimental treatments on the other. This we should do if we thought it the most important test, but sometimes we may be hoping to make equally important, or even more important, tests within the experimental treatments themselves, and so might be satisfied with something less than maximum precision for this test.

Orthogonal contrasts

As a biological research project develops, it becomes possible to make more sophisticated hypotheses. For economy of effort it is desirable to test each one independently, but include several in the same experiment, thereby making maximum use of the experimental material. In this way each unit can often contribute to answering several different questions. To see how to do this it is best to start thinking of the sum of squares for treatments. This represents all the information available on the variation among the treatments. We would like to divide it up into parts, each part representing a particular comparison, so that the S.S. for the individual comparisons sum to the total sum of squares for treatments. Then each subdivision would be independent in the statistical sense, and could be tested separately by means of an *F*-test. Such comparisons are said to be *orthogonal* to each other. Clearly we should expect to get most information if the ultimate subdivision produced as many parts as there are degrees of freedom available for the treatment S.S.

Let us first consider what orthogonal subdivisions are available for three treatments. Three treatments have two degrees of freedom, so we could expect to obtain answers to two null hypotheses of the form $A = B$ or $A - B = 0$. In statistical language $A - B$ is called a *contrast*. Let the treatment totals be a, b and c, each being a total of r replications, and let us decide that we want a versus (v.) b as one comparison. We can find the remaining comparison orthogonal to it by finding the remaining sums of squares in terms of a, b and c. The S.S. for a v. b will be

$$\frac{(a-b)^2}{2r}$$

Now the total sum of squares for treatments is

$$\frac{a^2+b^2+c^2}{r} - \frac{(a+b+c)^2}{3r}$$

so the remaining D.F. will have the variation provided by

$$\frac{a^2+b^2+c^2}{r} - \frac{(a+b+c)^2}{3r} - \frac{(a-b)^2}{2r}$$

$$= \frac{1}{6r}\{6a^2+6b^2+6c^2-2a^2-2b^2-2c^2-4ab-4ac-4bc-3a^2-3b^2+6ab\}$$

$$= \frac{1}{6r}\{a^2+b^2+4c^2+2ab-4ac-4bc\}$$

$$= \frac{1}{6r}\{(a^2+2ab+b^2)-4c(a+b)+4c^2\}$$

$$= \frac{1}{6r}\{(a+b)^2-4c(a+b)+(2c)^2\} = \frac{1}{6r}(a+b-2c)^2$$

which is the S.S. for the contrast $a+b-2c$ or equally the comparison of $\frac{1}{2}(a+b)$ and c. Thus for three treatments there is an orthogonal split into a v. b and $\frac{1}{2}(a+b)$ v. c, or in words the mean of a and b versus c, and their sums of squares add up to the total treatment sum of squares.

There is a neat way of writing this result in a sort of matrix notation which is very helpful for finding orthogonal contrasts and for testing

EXPERIMENTS WITH MORE THAN TWO TREATMENTS

orthogonality. We write a table of coefficients, i.e. the multipliers for each treatment total in each sum of squares expression:

	a	b	c
1st contrast	+1	−1	0
2nd contrast	+1	+1	−2

We can now see two properties: (1) that the coefficients for each contrast sum to zero, and (2) that the product of the two rows of coefficients sum to zero, i.e. $(+1)(+1)+(-1)(+1)+(0)(-2) = 0$. These are the rules for orthogonal contrasts and always apply. In the general case, calling the coefficients k, we have $\sum k_i = 0$ and $\sum k_i k_j = 0$ $(i \neq j)$. This applies over the whole series of contrasts, i.e. to be orthogonal the products of the coefficients of every pair of contrasts must sum to zero.

When the rules are stated in this way, the coefficients are those applied to individual values, but it applies directly to combining totals if each total is formed from the same number of replications. If, however, replication is not the same for each treatment, and the ks are applied to totals

$$k_1 r_1 + k_2 r_2 + \ldots = 0 \quad \text{and} \quad k_{11} k_{21} r_1 + k_{12} k_{22} r_2 \ldots = 0$$

i.e. applying k_1 to a total of r_1 values means the same as applying it to the r_1 values individually and adding these products together.

Now looking back at the algebra again, we see that for each sum of squares there was a difference to be squared, i.e. $a-b$ and $(a+b)-2c$, but also that this square had to be divided by a certain number as well as dividing by the number of replicates; $(a-b)^2$ was divided by 2, something we worked out on p. 25, while $(a+b-2c)^2$ was divided by 6. The rule is related to the argument on p. 30 that when the values are independent, variances (V) and consequently S.S. are additive, i.e.

$$V_{(x_1+x_2+\ldots)} = V_{x_1} + V_{x_2} + \ldots \quad \text{and that} \quad V_{cx} = c^2 V_x$$

so

$$V_{(c_1 x_1 \pm c_2 x_2 \pm \ldots)} = c_1^2 V_{x_1} + c_2^2 V_{x_2} + \ldots$$

and when $V_{x_1} = V_{x_2} = \ldots$, as in most experiments, this becomes

$$(c_1^2 + c_2^2 + \ldots) V_x$$

If we want S.S. in terms of single observations we must divide by $(c_1^2 + c_2^2 + \ldots)$. So $1^2 + (-1)^2 = 2$ and $1^2 + 1^2 + (-2)^2 = 6$, and this is the pattern throughout; the divisor for the sum of squares of a linear contrast is always the sum of the squares of the coefficients used to obtain the difference which was squared. In the general form the divisor for the ith contrast is $\sum k_i^2$.

Again it is best to work from the coefficients applied to individual values, but if there is the same replication (r) for each treatment then, considering the coefficients applied to the totals, the divisor becomes $r \sum k^2$ and with unequal replications ($r_1, r_2, r_3, \ldots$) it becomes $r_1 k_1^2 + r_2 k_2^2 + r_3 k_3^2 + \ldots$.

To see how this might be useful in biology, consider again the data from the primrose experiment on p. 55. Suppose now that the first habitat (h_0) was in a dry area, whilst the others (h_1 and h_2) were in wet areas. The botanist might well wish to test two separate null hypotheses: (1) When grown under the same environmental conditions there is no difference in stalk length between primroses obtained from different habitats within the wet area; and (2) There is no difference between those obtained from the dry area and those from the wet area. The experiment had equal replication (six-fold) of the habitats, so it is possible to use the treatment totals directly, and these would be set out under the treatment letters as in Table 5.5. Then we should work out the coefficients required to make the desired contrasts. The first is easy: if there is no difference between primroses from the two wet habitats $h_2 - h_1$ will equal zero, so we put $+1$ under h_2, -1 under h_1 and 0 under h_0, because the dry habitat does not enter into this argument. For the second null hypothesis we need a difference between the wet and dry habitats but, since there are more plants from the wet than the dry, obviously a simple difference $h_2 + h_1 - h_0$ would not equal zero if there was no difference. In general, to find appropriate coefficients in such cases, first determine the number of treatments (or individual values if there is unequal replication) on each side of the contrast (in this case 1 and 2), and determine the lowest common multiple of these numbers. The required coefficients for either side is simply the lowest common multiple divided by the number of treatments (or individual values in the case of unequal replication) on that side. By convention, the side containing the higher suffixes is given the positive sign. Then we should check that all the degrees of freedom have been used up, and that all the contrasts are mutually orthogonal. In the present case there are three treatments, so 2 D.F. for treatments, and we have made two contrasts so should not be able to find any more. $\sum k_i k_j = (0) \times (-2) + (-1) \times (+1) + (+1) \times (+1) = 0$ so our two contrasts are orthogonal to each other, and consequently their sums of squares will sum to the total sum of squares for treatments. It should be stressed that this is the

EXPERIMENTS WITH MORE THAN TWO TREATMENTS   75

sensible approach to the problem, because the biologist wants to test particular biological hypotheses and must start from these hypotheses; he must not take a standard statistical procedure and make hypotheses to suit the procedure.

Table 5.5—Linear contrasts

	h_0	h_1	h_2	$\sum k^2$
	469	582	538	
h_1 v. h_2	0	−1	+1	2
h_0 v. (h_1+h_2)	−2	+1	+1	6

Having checked for orthogonality we can now do a complete analysis of variance as in Table 5.6.

Table 5.6—Analysis of variance

Source of variation	D.F.	S.S.	M.S.	F
Within wet habitats (h_1 v. h_2)	1	161	161	1·20
Wet v. dry habitats [h_0 v. (h_1+h_2)]	1	920	920	6·87
Error	15	2013	134	
Total	17	3094		

Using the table of coefficients (Table 5.5) the sum of squares for h_1 v. h_2 is

$$\frac{\{(-1) \times 582 + (+1) \times 538\}^2}{2 \times 6} = \frac{(-44)^2}{12} = 161$$

and for h_0 v. (h_1+h_2) is

$$\frac{\{(-2) \times 469 + (+1) \times 582 + (+1) \times 538\}^2}{6 \times 6} = \frac{(+182)^2}{36} = 920$$

The total sum of squares was worked out on p. 58 as 3094, so the error S.S. by difference is 2013. Allowing for rounding inaccuracies this is the same as before (in fact working to two places of decimals the individual components of treatment S.S. are 161·33 + 920·11 = 1081·44, whilst the direct calculation gives 1081·44). When there are many orthogonal contrasts, it is a useful check to work out the total sum of squares for treatments both ways, to ensure that no arithmetical mistakes have been made.

Since each contrast has one degree of freedom, the M.S. is the same as the S.S. in each case, and each can be tested by dividing by the error M.S. to obtain F with $n_1 = 1$ and $n_2 = 15$. From Appendix Table II the F-value with these D.F. is 4·54 at $P \leqslant 0.05$ and 8·68 at $P \leqslant 0.01$. With our value of 6·87 for the wet v. dry contrast we can say that, if the null hypothesis that there is no difference in stalk length between primroses obtained from the wet and dry habitats is true, we have witnessed a very unlikely event and would prefer to believe that there is a difference. With an F-value of 1·20 for the other contrast, however, we would be quite content to go on believing that within the wet area the primroses form a homogeneous population.

Remembering that $F = t^2$ when there is only one degree of freedom for the treatment mean square, we could make the test another way. The values we calculated to represent the contrasts (-44 for h_1 v. h_2 and $+182$ for h_0 v. (h_1+h_2)) before squaring and dividing by the sums of squares of the coefficients, can be thought of as the total effect of the contrasts and can be tested with a t-test. Following the rules of combining variances (p. 73) the S.E. for a linear contrast when all the treatments have the same replication (r) and the same variance will be $\sqrt{r(\sum k^2)}$ M.S.$_E$. Thus for h_1 v. h_2 we have an effect of -44 with an S.E. of $\sqrt{(6 \times 2 \times 134)} = 40·1$ and

$$t = \frac{44}{40·1} = 1·10$$

(Note that we can omit the sign before 44 since we are testing if there is an effect, not whether it is positive or negative.) It will be seen that this test is exactly the same as testing a simple difference in a two-treatment experiment. This emphasizes that such an experiment is in no way a special case but is one of a general type, which has but one linear contrast available.

h_0 v. (h_1+h_2) has an effect of $+182$ and S.E. of $\sqrt{(6 \times 6 \times 134)} = 69·5$, so

$$t = \frac{182}{69·5} = 2·62$$

which is, as expected, significant at the 5 per cent level. Similarly t can be used to provide confidence limits for linear contrasts in the same way as for simple differences, e.g. the confidence interval for h_0 v. (h_1+h_2) is $\pm t_{15} \times$ the S.E. of h_0 v. (h_1+h_2) where t_{15} is the tabular value of t with 15 D.F. at whatever level of confidence is chosen. At 95 per cent confidence it becomes $\pm 2·13 \times 69·5 = \pm 148$ and we can say that we are 95 per cent confident that the true effect lies between $182-148$ and $182+148$ or 34 and 330. Total effects and their confidence limits can then be brought to the most appropriate units by multiplying or dividing by an appropriate constant.

In the present case we might wish to express the result in lengths of a single stalk in millimetres. The original data were means of single stalks and there were six such values added to give h_0, h_1 and h_2. The effect was $h_1 + h_2 - 2h_0$, so was a difference between $6+6$ and 2×6, i.e. between two totals of 12 values. To get back to a difference between single stalks we should divide by 12, giving an effect in terms of a single stalk of

$$\frac{182}{12} = 15\cdot2 \quad \text{with S.E. of} \quad \frac{69\cdot5}{12} = 5\cdot79$$

and confidence limits of

$$\frac{34}{12} \quad \text{and} \quad \frac{330}{12} = 2\cdot8 \text{ and } 27\cdot5$$

All could be converted to any other units by multiplying or dividing by an appropriate constant. Some workers refer to the effect just calculated as the mean effect and proceed to it directly by dividing the total effect by the sum of the positive coefficients multiplied by the number of replications, $(1+1) \times 6 = 12$, in this case. The S.E. of such a mean effect can also be calculated directly following the usual rules as

$$\sqrt{\frac{\sum k^2 \times \text{M.S.}_E}{r\{\sum(+k)\}^2}}$$

where $\sum(+k)$ means the sum of all coefficients with a positive sign which can also be written $\frac{1}{2}\sum|k|$ where $\sum|k|$ means the sum of all coefficients, ignoring the sign. In the present case the S.E. would be

$$\sqrt{\frac{6 \times 134}{6 \times 2 \times 2}} = 5\cdot79$$

as before. However, there is room for debate as to what is the best definition of a mean effect, and the words should not be used without qualification. It is better to proceed from the basic biological point of view and arrive at an answer in simple biological terms. Here we should say that our best estimate of the difference in length of stalk between primroses from the dry habitat and those from the wet habitat is $15\cdot2 \pm 5\cdot79$ mm per stalk, something everyone can understand.

If there is unequal replication, the total effect of a linear contrast and its standard error are still easily calculated. As given earlier the effect is $\sum kT$ where the Ts are the treatment totals and the S.E. is $\sqrt{\text{M.S.}_E(r_1 k_1^2 + r_2 k_2^2 + \ldots)}$. To convert to mean contrasts we divide by $\frac{1}{2}\sum|rk|$.

The number of sets of orthogonal contrasts possible for three treatments is limitless, but once one member of the set is chosen, the other is fixed because there will be only one contrast which is orthogonal to it. For example if we chose a first contrast of

	a_0	a_1	a_2	
	-4	-1	$+5$	the second contrast must be
	$+2$	-3	$+1$	

In general, if there are p treatments, we can divide the sums of squares into sets of $p-1$ components to represent $p-1$ contrasts. Once we have decided $p-2$ of them, the remaining one is also decided. Choosing the contrasts to test is the essence of experimental design, and the decision should be completely controlled by the null hypotheses, and therefore made before the experiment starts. To look at the data and decide from them which contrasts to *test* is simply a form of self-deception. This should not deter the experimenter from gaining further experience from his data. If he sees that some contrasts that he had not thought of testing look large, the sensible thing to do is to make more experiments in which they form part of the null hypothesis, so that a deliberate attempt is made to test them.

The two orthogonal contrasts used for the primroses are the ones most commonly used for three treatments in biology. Setting out the coefficients again as

	a_0	a_1	a_2
1st contrast	-2	$+1$	$+1$
2nd contrast	0	-1	$+1$

we can see several uses. For example, if we have two formulations of a chemical substance as a_1 and a_2, and no chemical applied as a_0, the first contrast tells us if the chemical has an effect, and the second if the method of formulation matters. Again, we might think that adding an antibiotic to the food of an animal will affect its growth rate, but have little idea of how much to add. Making a_0 no antibiotic, a_1 a small quantity, and a_2 a larger quantity, the first contrast tells us if the antibiotic has an effect, and the second if the quantity applied matters. At a later stage in such a research project we may get the idea that there is a gradual increase in growth as the quantity of antibiotic is increased, and would wish to test if the increase is linear. Taking a_0 as without antibiotic, a_1 as 2 mg/kg food, and a_2 as 4 mg/kg, pictorially we expect figure 5.1. Geometry tells us that if the effect is perfectly linear,

EXPERIMENTS WITH MORE THAN TWO TREATMENTS

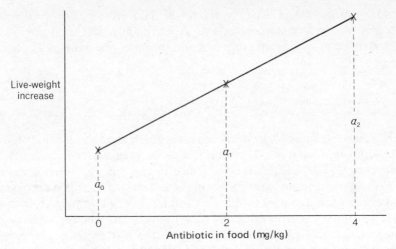

Figure 5.1—Effect of an antibiotic on live-weight increase.

$a_1 = \frac{1}{2}(a_0 + a_2)$ or $a_0 + a_2 - 2a_1 = 0$ so testing the first contrast of

	a_0	a_1	a_2
1st contrast	+1	−2	+1
2nd contrast	−1	0	+1

will tell us if this is reasonably true. If it is significant, $a_0 + a_2 - 2a_1$ does not equal zero unless an unlikely event has occurred, so a_1 cannot be said to lie on the line joining a_0 and a_2. We could say that there was a significant deviation from linearity over the range tested, and design more experiments to investigate the relationship more fully. The second contrast is the linear effect. If we put it in units of a single animal, it would be an estimate of the increase due to 4 mg/kg food; if the first contrast was not significant, it would be reasonable to divide this effect by 4, and thus provide an estimate of live-weight increase per mg antibiotic/kg food over the whole range of 0–4 mg/kg.

When we consider more treatments, we start to get several possible useful sets of orthogonal contrasts. Again we must decide which set to use according to the hypotheses, not according to the data. The first obvious set with four treatments will be an extension of the one we used with three.

Treatments	A	B	C	D	$\sum k^2$	
1st contrast	−1	+1	0	0	2	= A v. B
2nd contrast	−1	−1	+2	0	6	= C v. mean of (A+B)
3rd contrast	−1	−1	−1	+3	12	= D v. mean of (A+B+C)

Each sums to zero in itself, 1st × 2nd = 0, 1st × 3rd = 0, 2nd × 3rd = 0, so they are orthogonal contrasts as expected. This sort of system can be extended to any number of treatments. Another possibility is

Treatment	A	B	C	D	$\sum k^2$
1st contrast	−1	−1	+1	+1	4
2nd contrast	−1	+1	0	0	2
3rd contrast	0	0	−1	+1	2

The sums and sums of products are zero which shows again that these are orthogonal contrasts. This set can be useful when we have to compare two things, each of which can be treated in two different sorts of ways, e.g. in an agricultural experiment concerned with soil fertility we might ask: does grassland lead to different soil fertility from arable cropping. But we cannot just grow grass. We have to use it, and whether we mow it or graze it, could affect the comparison with arable cropping. Likewise arable cropping might give a different effect depending on whether the crop residues were returned to the soil or not, and both are very reasonable interpretations of arable cropping. So we would make A mown grassland, B grazed grassland, C arable cropping with return of residues, and D arable cropping without return of residues. Then the first contrast shows if grassland on average differs from arable cropping on average; the second tells us if the grassland effect is dependent on management, and the third if the arable effect is dependent on return of residues.

CHAPTER SIX

FACTORIAL ARRANGEMENT OF TREATMENTS

Let us suppose that a biochemist is studying the function of amino acids in determining the growth of young chicks and that he has come to the point where he considers that one or both of methionine and arginine might be important. He would wish to design an experiment to see if the addition of extra methionine to the diet affects growth, if the addition of extra arginine affects growth, and also if these amino acids act independently. He would therefore have three null hypotheses which he would like to test in a single experiment. Using the symbols m_0 for no added methionine, m_1 for added methionine, a_0 for no added arginine and a_1 for added arginine, he needs to test $a_1 - a_0$, $m_1 - m_0$ and whether a_1 and m_1 given together have the same total effect as they would if given singly. Suppose he chooses 24 chicks and at random allocates six to each of the treatments a_0m_0, a_0m_1, a_1m_0 and a_1m_1 and obtains the treatment totals of live-weight gain shown in Table 6.1. Now to test $a_1 - a_0$ he could use the first contrast, i.e. the total of all the chicks which had supplementary arginine minus the total of all those that did not. Some might question if

Table 6.1—Live-weight gain and orthogonal contrasts

Treatments	a_0m_0	a_1m_0	a_0m_1	a_1m_1
Treatment totals	832	853	881	966
1st contrast	-1	$+1$	-1	$+1$
2nd contrast	-1	-1	$+1$	$+1$
3rd contrast	$+1$	-1	-1	$+1$

this is really what he wanted, because it is not a pure comparison of $a_1 - a_0$ alone, but an average of $a_1m_0 - a_0m_0$ and $a_1m_1 - a_0m_1$, i.e. an average over the two levels of methionine. This is no difficulty in most biological experiments, because the chicks have to have some methionine and there will be some in the standard diet in any case. It is implicit in all null hypotheses that the result is being assessed within the bounds set by the population from which the experimental samples were drawn, and in a case such as this these bounds obviously include the amount of methionine in the standard diet. Thus it is just as much biological sense to talk of the average effect of arginine on the two diets with two defined methionine contents as it is to talk of its effect on this single standard diet, or any other diet. We therefore say that the first contrast gives the average effect of arginine, often referred to as the *main effect* or *straight effect* in some statistical jargons, and whenever there are only two levels, is given the capital letter corresponding to the lower-case letter used to denote the treatments. In this case we would use A to denote this main effect.

Likewise the second contrast is the main effect M of methionine. It is $(a_0m_1 - a_0m_0) + (a_1m_1 - a_1m_0)$, i.e. an average effect over the two levels of arginine. If we sum the products of the coefficients we get $(-1)(-1) + (+1)(-1) + (-1)(+1) + (+1)(+1) = 0$; so these two contrasts are orthogonal, and a little thought will reveal that the only contrast orthogonal to both of them has as coefficients the products of the coefficients of these two, i.e. $+1, -1, -1, +1$ from above.

Now this contrast is $(a_0m_0 + a_1m_1) - (a_0m_1 + a_1m_0)$ and it tells us if the two amino acids act independently because, on the positive side of the contrast, we have one dose of arginine and one dose of methionine given to the same chicks (a_1m_1), whilst on the negative side we have single doses of each given to different chicks; if the effects of the two are independent we would expect this contrast to be zero.

This contrast is called the *interaction* between the two factors and given the symbols $A \times M$ or just AM. Another interpretation may be seen by setting it out in the form $(a_1m_1 - a_0m_1) - (a_1m_0 - a_0m_0)$. The first bracket contains the effect of a when m is present, and the second the effect of a when m is absent. Thus the contrast is the difference in the effect of a which is brought about by giving m.

These effects can be tested by producing sums of squares by squaring the effects and dividing by the sums of squares of the coefficients and the number of replicates, as in the previous example, and tested with an F-test, or the actual total effects can be tested with a t-test as before.

$$\text{Thus S.S. of } A = \frac{(-1 \times 832 + 1 \times 853 - 1 \times 881 + 1 \times 966)^2}{4 \times 6} = \frac{106^2}{24} = 468 \cdot 2$$

$$\text{of } M = \frac{(-1 \times 832 - 1 \times 853 + 1 \times 881 + 1 \times 966)^2}{4 \times 6} = \frac{162^2}{24} = 1093 \cdot 5$$

$$\text{and of } AM = \frac{(+1 \times 832 - 1 \times 853 - 1 \times 881 + 1 \times 966)^2}{4 \times 6} = \frac{64^2}{24} = 170 \cdot 7$$

FACTORIAL ARRANGEMENT OF TREATMENTS

These can now be entered in the analysis of variance table (Table 6.2): each has 1 D.F. and M.S. = S.S./1 = S.S., and they can be tested with an F-test: for M, F is just over 5 and is therefore significant at the 5 per cent level; for A, F is less than $2\frac{1}{2}$ so is not significant, and for AM it is less than 1 so is not significant. If we stick to the null hypotheses in this type of experiment there is no possibility of confusion; in this case our biochemist would be entitled to say that he has good evidence for believing

Table 6.2—Analysis of variance

Sources of variation	D.F.	S.S.	M.S.	F
Arginine (A)	1	468		2·38
Methionine (M)	1	1094		5·57
Interaction (AM)	1	171		0·87
Error	20	3926	196·3	
Total	23	5659		

that adding methionine to the diet affects the growth rate of growing chicks, and he could give confidence limits for the extra growth expected. He could also say that when he tested it he could not find sufficient evidence to convince him that arginine had any effect on average, nor that these two amino acids acted other than independently. In this form of argument both the biological questions and the statistical tests are independent. However, troubles arise when scientists try to use the values obtained for further "interpretation" of the results. In doing so they are usually contravening the rule of basing tests on null hypotheses, and are selecting additional tests because of the appearance of the data. Such "interpretations" should be used only as a guide to further experimentation, not published with S.E.s as if they had been obtained from an experiment designed for that purpose. This trouble most commonly occurs in "interpreting" the interaction. If it had been significant the test would have told us that $(a_1 m_1 + a_0 m_0)$ does not equal $(a_0 m_1 + a_1 m_0)$; it would not tell us $a_1 m_1 \neq a_0 m_1$ whilst $a_1 m_0 = a_0 m_0$, which is the sort of question we might like to ask if we knew that the interaction existed. Although the treatment combinations required to test these null hypotheses might be the same as in our experiment, we shall see later that the desirable replication could be different. We have already seen in Chapter 5 that the orthogonal contrasts would be different, indeed the only contrast orthogonal to both $(a_1 m_1 - a_0 m_1)$ and $(a_1 m_0 - a_0 m_0)$ is $(a_1 m_1 + a_0 m_1) - (a_1 m_0 + a_0 m_1)$, i.e. the effect M.

χ^2-test

The use of F-tests or t-tests in this type of design demands that the data

FACTORIAL ARRANGEMENT OF TREATMENTS

are normally distributed, and there are as yet no non-parametric procedures which explore such situations very well. There is, however, one test which works like an interaction test and which might be worth mentioning here—that is the χ^2-test. It is appropriate when the data are counts.

For example, a physiologist wishing to investigate the effects of pre-natal nutrition on twinning in sheep might take a flock of ewes and allocate them at random to two treatments: a high plane of nutrition and a low plane. The only measures available for testing any effects might be the number of ewes which had twins, and the number that had singles in each group:

	Twins	Singles	Total
High plane	36	84	120
Low plane	26	97	123
Total	62	181	243

It is worth remarking here that in such an experiment we would usually start with the same number of animals on each treatment, but losses during the experiment often mean that we end up with unequal replication.

Now if the plane of nutrition had no effect on twinning, we should expect the proportion of twins to singles to be the same for each plane, i.e. since on average $\frac{62}{243}$ have twins and $\frac{181}{243}$ have singles, of the 120 on the high plane we would expect $\frac{62}{243} \times 120$ to have twins and $\frac{181}{243} \times 120$ to have singles. Likewise for the low plane we expect $\frac{62}{243} \times 123$ to have twins and $\frac{181}{243} \times 123$ to have singles or, in full, if our null hypothesis that nutrition has no effect on twinning is true, we expect:

	Twins	Singles	Total
High plane	30.6	89.4	120
Low plane	31.4	91.6	123
Total	62	181	243

To test if our observed values are so far from those expected if the null hypothesis is true that we prefer to believe that it is not true, we first calculate the deviations of the observed values (O) from the expected values (E) giving:

	Twins	Singles	Total
High plane	+5.4	−5.4	0
Low plane	−5.4	+5.4	0
Total	0	0	0

$\sum \frac{(O-E)^2}{E}$ follows approximately a well-known statistical distribution called χ^2.

Tables are given in Fisher and Yates (1963) and a few values in Appendix Table III. In the present case

$$\sum \frac{(O-E)^2}{E} = \frac{(+5\cdot4)^2}{30\cdot6} + \frac{(-5\cdot4)^2}{89\cdot4} + \frac{(-5\cdot4)^2}{31\cdot4} + \frac{(+5\cdot4)^2}{91\cdot6} = 2\cdot53.$$

χ^2-tables are two-dimensional, the value depending on the degrees of freedom of the deviations being tested, and the level of probability required, and are given in the same form as t and F, namely the largest value that can be obtained at that level of probability if the null hypothesis is true. Degrees of freedom for the test can be worked out in the same way as we did in the previous case. There are four observations, so 3 D.F. in total; 1 D.F. represents the difference between the numbers of ewes on the high and low planes, which was under the experimenter's control so does not require testing; 1 D.F. represents the difference between the numbers of ewes having twins and singles, which again answers no hypothesis in this case; and the other 1 D.F. represents the "interaction" effect we are seeking. Remembering the argument that degrees of freedom tell us the number of independent deviations, it is thus not surprising to see that in the present case all the deviations are of the same magnitude (i.e. all + or −5·4). Referring to the χ^2-table under 1 D.F. we see that 2·53 has a probability greater than 0·10, so most workers would see no reason for disputing the null hypothesis, and therefore no reason to believe that pre-natal nutrition affects twinning in the population of sheep investigated.

The method used for the calculations here applies quite generally, no matter how many levels each factor has, but in the special case of a 2×2 situation there is a simpler method. Calling the observed values in the table a, b, c, d, as follows

	Twins	Singles	Total
High plane	a	b	$a+b$
Low plane	c	d	$c+d$
Total	$a+c$	$b+d$	$a+b+c+d$

$$\chi^2 = \frac{(ad-bc)^2(a+b+c+d)}{(a+b)(c+d)(a+c)(b+d)} = 2\cdot51 \text{ in this case, a more accurate estimate}$$

since the only rounding to be done was at the end of the calculation, whereas in the previous method $\pm 5\cdot4$ was not exact and had to be squared.

This method is used only with observations which are counts. If the expected number in any of the combinations is very small, $\sum[(O-E)^2/E]$ does not follow the χ^2-distribution very closely so, if any cell has an expected number of 5 or less, the usual advice is: in the 2×2 situation use Yates' correction (Yates, 1934); in any other situation combine rows or columns until no expectation is less than 5 (of course this means discarding some of the null hypotheses, so perhaps better advice would be to repeat the experiment, with more observations or a better design).

Advantages of factorial experiments

The experiment on amino acids as originally designed and analysed is one of a very important and useful class of design called *factorial designs*. Yates (1937) gave an early account of them. They are appropriate when there are two or more sorts of treatments to be applied in different amounts which can be applied together as well as alone. In statistical terms, the sort of treatment is called a *factor*, and the amount applied is called the *level* of the factor. In our example, we had two factors (a and m) each at two levels (some and none), but factorial experiments can have as many factors as we like, and each factor as many levels as we like; all factors do not have to have the same number of levels, and levels need not be actual quantities in numerical terms, e.g. in a test of several cultivars we might have two factors, fertilizer application and cultivar, where the fertilizer would have numerical levels, but the levels of cultivar would be the different genotypes, A, B, C, D, say. The essence of the factorial design is that the treatments are made up of all possible combinations of the levels of the different factors, so if there are two factors, one at p levels and the other at q levels, there will be pq treatments, and so on. But probably the most useful factorial experiments for biologists are those in which all factors have two levels; they are referred to 2^n designs where n denotes the number of factors (2, of course, denotes the number of levels of each) and it is worth looking into these in more detail.

We might ask what is their particular advantage over simple two-treatment experiments like the fungicide experiment in Chapter 2. There are three real advantages.

(1) Suppose that in the amino acid example only 24 chicks were available; if simple two-treatment experiments were used, 12 would be available for testing A and 12 for testing M, so we should allocate 6 to a_0 and 6 to a_1 in one experiment, and 6 to m_0 and 6 to m_1 in another experiment.

Each experiment would have an analysis of variance of

	D.F.
Treatments	1
Error	10
Total	11

whereas by using the factorial arrangement we had

	D.F.
Treatments	3
Error	20
Total	23

with a very much better estimate of error based on 20 D.F. than on 10 D.F., and consequently the possibility of detecting smaller true differences for that reason. Furthermore, when testing treatment effects such as A, the replication in the first case is only 6-fold, so the S.E. would be $\sqrt{(2 M.S./6)}$. In the second case the effective replication is 12, because half the chicks have a_0 and half have a_1, so the S.E. would be $\sqrt{(2 M.S./12)}$. Hence a smaller difference will be shown significant at any given level of probability in the second case than in the first.

(2) The second advantage is this: By making the two simple experiments we should never know if there was an interaction or not, whereas in the factorial experiment we can always test the interaction. Thus we get more information for the same amount of work.

(3) The third advantage is shown in many fields of research, where in any substantial research problem there are bound to be several factors involved. If we wanted to introduce a new crop to an area, we should be interested in finding the best seed rate, the best row-spacing, the best fertilizer treatment, etc. Suppose we started with an experiment on seed rate. The crop has got to have some sort of row-spacing, and some fertilizer, so we would have to guess these for the first experiment. Then, knowing a suitable seed rate from that experiment, we might test row spacing. Suppose we found the best was widely different from the one used in the seed-rate trial, then we must go back and do the seed-rate trial all over again, using this row-spacing, and so on, involving continuous repetition, whereas a factorial experiment could take care of all possibilities in a single trial. Indeed, at almost every stage in biological research we find doubts creeping in as to whether we have set the best hypotheses to test, and we wonder whether some other factor might affect the response to the factors under consideration. Factorial design allows the experimenter to include these other factors with little more work and, if they

are important, he is saved from having his work discredited later on when someone finds that his conclusions are true only under certain restricted levels of these other factors. If the other factors are not important determinants of the response he has measured, he still benefits by being able to specify the responses over a wider range of conditions.

Designing a factorial experiment

In designing a factorial experiment we first have to work out all the treatment combinations. In 2^n designs there are two recognized ways of denoting the treatment for each experimental item.

(1) The way we used earlier of giving the level of each factor, i.e. a_0m_0, a_1m_0, a_0m_1 and a_1m_1.

(2) We can give the factor letter whenever the second level is present, it being understood that the first level of any factor is present if no letter for that factor appears. By convention the items which receive the first level of all factors are given the symbol (1). This is a convenience we shall see later when we need to multiply symbols together; if we used 0 for these levels we should have to redefine the algebra. Thus we could have written (1), a, m, am to give us the same treatments as before. In this convention it is usual to use small letters to denote treatments and capitals to denote effects as before. This latter method is very easy to work with and is quite descriptive of the treatments if the levels are presences and absences, as they often are in nutrient trials or feeding trials. It calls for more care if both levels represent positive things, e.g. if we had species of grass as a factor, where a_0 represented cocksfoot, and a_1 ryegrass in all combinations with management treatments of m_0 = mown and m_1 = grazed. Then we have to think carefully to see that (1) is mown cocksfoot, a is mown ryegrass, m is grazed cocksfoot, and am grazed ryegrass; setting these out as a_0m_0, a_1m_0, a_0m_1, a_1m_1 would be much more obvious. We should really become familiar with both methods and use whichever seems best for each experiment. Whichever method is used, the complete set of combinations is then allocated to the experimental items by the process of randomization in one operation.

Let us suppose that we have three factors A, B and C, each at two levels (i.e. a_0, a_1; b_0, b_1; c_0, c_1) to be tested, so we need a 2^3 experiment, i.e. eight treatments in all. Using symbols to describe both levels, the best way of ensuring that we miss nothing is to write down the first factor at its two levels 2^{n-1} times, in order, i.e.

$$a_0 \quad a_1 \quad a_0 \quad a_1 \quad a_0 \quad a_1 \quad a_0 \quad a_1$$

FACTORIAL ARRANGEMENT OF TREATMENTS

Then write beside these the second factor letter at both levels, writing each level twice consecutively, i.e.

$$a_0b_0 \quad a_1b_0 \quad a_0b_1 \quad a_1b_1 \quad a_0b_0 \quad a_1b_0 \quad a_0b_1 \quad a_1b_1$$

Then write beside these the third factor letter, writing each level four times consecutively, i.e.

$$a_0b_0c_0 \quad a_1b_0c_0 \quad a_0b_1c_0 \quad a_1b_1c_0 \quad a_0b_0c_1 \quad a_1b_0c_1 \quad a_0b_1c_1 \quad a_1b_1c_1$$

If there was a fourth factor, it would be written in blocks of eight and so on.

Building up the treatment combinations using the other system is most easily done by writing (1) first; then multiply it by the first factor letter, $a \times 1 = a$; then multiply all combinations written down so far by the second factor letter, $(1) \times b = b$, $a \times b = ab$, and continue with the next factor letter and so on giving

$$\begin{array}{cccccccc} (1) & a & b & ab & c & ac & bc & abc \\ \hline \times\, a & & & & & & & \\ \hline \times\, b & & & & & & & \\ \hline \times\, c & & & & & & & \end{array}$$

We should then consider how many replications are required. Here we should be very sure what null hypotheses are to be tested or what contrasts are to be given confidence limits. We have seen already that if the null hypotheses to be tested are the usual factorial ones, the main effect of A is zero, the main effect of B is zero ..., the interaction AB is zero ..., and so on, the design itself gives internal replication, there being 2^{n-1} replications of a_0 and 2^{n-1} of a_1 in every complete replicate. In the 2^3 situation, we might be satisfied with three complete replicates, which give therefore $3 \times 2^2 = 12$ replications and an S.E. of $\sqrt{(\frac{2}{12}\text{M.S.}_E)}$ for the mean effect A. Disappointment will arise if the experimenter really had null hypotheses like $a_0b_0c_1 = a_0b_0c_0$, which, of course, might require the same combinations, but such a contrast would have but 3 replications and an S.E. of $\sqrt{(\frac{2}{3}\text{M.S.}_E)}$, a very different matter. He who starts by setting the usual null hypotheses and complains that he cannot "interpret" the interactions because of insufficient precision for making tests of this sort, deserves little sympathy, because he is obviously making hypotheses from the data before him, and trying to use the same data to "test" his hypotheses.

Analysis of a 2^3 experiment

The experiment to be analysed here had four replications in a completely randomized design.

To do the analysis we need to get the correction factor and the total sum of squares

$$\sum x^2 - \frac{(\sum x)^2}{n}$$

in exactly the same way as we have for previous examples. In addition, we can get the sum of squares for each factor effect and each interaction, get the error S.S. by difference, and test all the factors and interactions with F-tests; or we can get the total treatment sum of squares, the error S.S. again by difference, and test all the factors and their interactions with t-tests. Whichever we do, we need to calculate an estimate of the effects of each factor and each interaction. There are several ways of doing this.

One useful method makes direct use of a table of coefficients, such as we used for getting other orthogonal contrasts. We write down the treatments in the same order as we wrote them down when finding all the combinations, and write the total for each treatment combination under

	$a_0b_0c_0$	$a_1b_0c_0$	$a_0b_1c_0$	$a_1b_1c_0$	$a_0b_0c_1$	$a_1b_0c_1$	$a_0b_1c_1$	$a_1b_1c_1$
or	(1)	a	b	ab	c	ac	bc	abc
Total	41·6	41·4	40·2	44·7	38·3	44·2	34·6	45·3
A	−	+	−	+	−	+	−	+
B	−	−	+	+	−	−	+	+
AB	+	−	−	+	+	−	−	+
C	−	−	−	−	+	+	+	+
AC	+	−	+	−	−	+	−	+
BC	+	+	−	−	−	−	+	+
ABC	−	+	+	−	+	−	−	+

the appropriate letters. Then to find which must be added and which subtracted for each effect, start with the first main effect A, and write + wherever a_1 or a is present, and − wherever a_0 is present or a is absent. Since all the coefficients are either $+1$ or -1, we can omit the 1s and simply write the signs. We then do the same for the second main effect B. To find the signs for AB, simply multiply these two rows of signs ($- \times - = +$; $+ \times - = -$ and so on). Next we produce a row for C in the same way as we did for A and B, and then multiply this row by the A row to get AC, by the B row to get BC, and by the AB row to get ABC.

Now for each effect we simply add up all the treatment totals, giving

FACTORIAL ARRANGEMENT OF TREATMENTS 91

each the sign detailed in the row for that effect, e.g.

$$A = -41\cdot6+41\cdot4-40\cdot2+44\cdot7-38\cdot3+44\cdot2-34\cdot6+45\cdot3 = +20\cdot9$$
$$ABC = -41\cdot6+41\cdot4+40\cdot2-44\cdot7+38\cdot3-44\cdot2-34\cdot6+45\cdot3 = +\ 0\cdot1$$

The remainder are:

$$B = -0\cdot7;\quad AB = +9\cdot5;\quad C = -5\cdot5;\quad AC = +12\cdot3;\quad BC = -4\cdot5$$

This concept of a coefficients table, or plus-and-minus table as it is often called, is very useful for showing exactly what is being compared in a complex experiment, and is essential for designing the more complicated experiments that we shall deal with later.

However, for the present purpose of working out all the effects there is a simpler and more mechanical method given by Yates (1937). We write down the treatment combinations in a column in the same systematic order which we have been using, and write the treatment totals in an adjacent column.

Treatment	Treatment totals	1	2	3	Total effect
(1)	41·6	83·0	167·9	330·3	Total
a	41·4	84·9	162·4	+20·9	A
b	40·2	82·5	+4·3	−0·7	B
ab	44·7	79·9	+16·6	+9·5	AB
c	38·3	−0·2	+1·9	−5·5	C
ac	44·2	+4·5	−2·6	+12·3	AC
bc	34·6	+5·9	+4·7	−4·5	BC
abc	45·3	+10·7	+4·8	+0·1	ABC

Then add the first two yields together and write the answer in the first row of the next column (call this column 1), then add the 3rd and 4th and write the answer in the second row of column 1. Add the 5th and 6th, and write in the third row, and the 7th and 8th, and write in the fourth row. Now subtract the 1st from the 2nd and write in the fifth row, subtract the 3rd from the 4th and write in the sixth row, 6th minus 5th in the seventh row, and 8th minus 7th in the eighth and last row. Now repeat this process of getting sums and differences of pairs to produce column 2 from column 1, in exactly the same way as column 1 was obtained from the treatment total column, and repeat just once more to produce column 3 from column 2. Now the values in column 3 are the total effects that we calculated previously; in the first row we have the grand total, in the second the effect of A, in the third the effect of B, in the fourth the effect of AB, and so on. We can complete our table by writing a column naming the effects by simply writing the capital letter corresponding to the lower-case letter in the first column.

This method is perfectly general for 2^n experiments, the number of columns of sums and differences to be worked out being n, i.e. there would have been 2 columns only for our $a \times m$ experiment which was 2^2, three for the 2^3, four for 2^4, and so on. Perhaps, to convince ourselves we might look at the algebra in a 2^2 experiment. Let us denote the individual treatment totals themselves by the small letters; we then have

Treatment total	1	2	Effect
(1)	$a+(1)$	$ab+b+a+(1)$	Total
a	$ab+b$	$ab-b+a-(1)$	A
b	$a-(1)$	$ab+b-a-(1)$	B
ab	$ab-b$	$ab-b-a+(1)$	AB

which checks with the contrasts we worked out earlier (p. 81).

The second method can be easily programmed on a computer. It is the quickest method on a calculating machine if we wish to work out *all* the effects, which is usually the case with 2^2 or 2^3, but when n is much greater than 3 we might consider that many of the higher-order interactions are nothing other than natural variation and would not want to calculate them, so the first method would be used, simply picking out the few appropriate rows from the total available. In both the methods we have to be very careful in computing, and the figures should be checked as well as possible. This is best done when all the total effects have been calculated by computing the total treatment sum of squares from these effects and seeing that it is the same as that obtained by squaring the individual treatment totals, dividing by the number of replications, and subtracting the correction factor. In our case the total treatment sum of squares is

$$\tfrac{1}{4} \times (41 \cdot 6^2 + 41 \cdot 4^2 + 40 \cdot 2^2 + 44 \cdot 7^2 + 38 \cdot 3^2 + 44 \cdot 2^2 + 34 \cdot 6^2 + 45 \cdot 3^2)$$
$$- 3409 \cdot 32 = 22 \cdot 79$$

The individual sums of squares for effects are the individual effects squared and divided by $r \sum k^2$ or $4 \times 8 = 32$. (It is a good check on the method that in the 2^n design, since all the ks are 1, and all items enter into each effect, the divisor is always the total number of items.)

FACTORIAL ARRANGEMENT OF TREATMENTS 93

S.S. for
$$A = \tfrac{1}{32} \times (+20\cdot9)^2 = 13\cdot65$$
$$B = \tfrac{1}{32} \times (-0\cdot7)^2 = 0\cdot02$$
$$AB = \tfrac{1}{32} \times (+9\cdot5)^2 = 2\cdot82$$
$$C = \tfrac{1}{32} \times (-5\cdot5)^2 = 0\cdot95$$
$$AC = \tfrac{1}{32} \times (+12\cdot3)^2 = 4\cdot73$$
$$BC = \tfrac{1}{32} \times (-4\cdot5)^2 = 0\cdot63$$
$$ABC = \tfrac{1}{32} \times (+0\cdot1)^2 = 0\cdot00$$
$$\text{Total} \qquad 22\cdot80$$

Allowing for rounding errors this is the same as we obtained by the other method. If we simply wanted to check the effects without recording the individual sums of squares, we could have done so by simply summing the squares of the effects and dividing the sum by 32, i.e.

$$\tfrac{1}{32} \times [(+20\cdot9)^2 + (-0\cdot7)^2 + \ldots + (+0\cdot1)^2] = 22\cdot79 \quad \text{as before}$$

If we are going to use F-tests we should set out a full analysis of variance table as Table 6.3.

Table 6.3—Analysis of variance

Source of variation	D.F.	S.S.	M.S.	F
A	1	13·65		22·38***
B	1	0·02		<1
AB	1	2·82		4·62*
C	1	0·95		1·56
AC	1	4·73		7·75*
BC	1	0·63		1·03
ABC	1	0·00		0
Error	24	14·64	0·610	
Total	31	37·44		

Reference to F with 1 and 24 D.F. in Appendix Table II (p. 227) shows that the value for A exceeds the 0·1 per cent value and those for AB and AC both exceed the 5 per cent value; so we should be convinced that on average A has an effect and that neither A and B nor A and C act independently, but be quite prepared to go on believing that neither B nor C has any effect on average and that there is no interaction effect of BC or ABC.

The other way we could test the effects is to use the t-test. In this case we require an abbreviated analysis of variance (Table 6.4).

Table 6.4—Analysis of variance

Sources of variation	D.F.	S.S.	M.S.
Treatments	7	22·79	
Error	24	14·65	0·610
Total	31	37·44	

From this we obtain the error M.S. from which to calculate the S.E.s of the effects. Since each total effect was obtained by multiplying totals of r plots by the coefficients k, the S.E. will be

$$\sqrt{(M.S._E \times r \times \sum k^2)}$$

and any effect greater than $t \times$ S.E. will be significant at that level of t. In our case

$$S.E. = \sqrt{(0·610 \times 4 \times 8)} = \sqrt{19·52} = 4·42$$

and with 24 D.F.

$$t \times 4·42 = 2·06 \times 4·42 = 9·11 \quad \text{at} \quad P = 0·05$$
$$2·80 \times 4·42 = 12·38 \quad \text{at} \quad P = 0·01$$
$$3·74 \times 4·42 = 16·53 \quad \text{at} \quad P = 0·001$$

Referring to the total effects set out on p. 91 we see that A is greater than 16·53 and so very highly significant; AB and AC are both greater than 9·11 but not greater than 12·38, so are significant at the 5 per cent level; whilst all the rest are less than 9·11, and so for them the results do not dispute the null hypotheses.

When displaying the results to the rest of the world, we need only show values for effects that we believe to be true. If only the main effect A was significant, all the useful information would be contained in a statement of the effect of A and its standard error. It is normal to put the effect into units which are easily understood by other biologists. This might be just the difference between two means, i.e. others could easily grasp the results if told the difference between the means of all those that received a and those which did not, as they would then expect this sort of difference to apply to any similar single member of the population; and if they knew the S.E. they could also set confidence limits to this value. Following this argument, if only A were significant in this example, we could obtain this form of mean effect directly by adding all the values for a_1 and dividing by the number of values, i.e. 4×4, because there were four treatment combinations containing a_1, each with four replications, giving the mean for a_1 as $\frac{1}{16} \times (41\cdot4 + 44\cdot7 + 44\cdot2 + 45\cdot3) = 10\cdot98$. Likewise the mean for a_0 is $\frac{1}{16} \times (41\cdot6 + 40\cdot2 + 38\cdot3 + 34\cdot6) = 9\cdot67$. The S.E. for each of these means is $\pm\sqrt{(\frac{1}{16}\text{M.S.}_E)}$ following the usual rules, so we could display the results as

	a_0	a_1	S.E.
Mean	9·67	10·98	±0·195

Alternatively, we could give the difference between these means and the S.E. of this difference, i.e. $\sqrt{(\frac{2}{16}\text{M.S.}_E)}$ and simply state:

the mean effect of A is $+1\cdot31 \pm 0\cdot276$

If the experimental means are not the usually understood units, they can be converted to any other units by simply multiplying the effect and its standard error by an appropriate constant. For example, if our data had been the growth of an organism over a period of eight days in grams, we could divide through by 8 and multiply by 1000 and say:

A increases the growth by $164 \pm 34\cdot5$ mg/day

If, however, an interaction is significant, the average effect of any of the factors involved will be of less interest, because it will then be known that this factor does not have a stable effect over a wide range of conditions, but that its effect changes with changes in level of other factors. If we are unlikely to be using it with the average level of the other factors, its average effect is not worth knowing. Indeed, in research, the ultimate object is often to find the best combination of levels of the various factors. So if two-factor interactions are significant, the results are best displayed as two-way tables. In the present case AB and AC are both significant, so two tables, one showing means for all combinations of a and b averaged over both levels of c, and the other showing means for all combinations of a and c averaged over both levels of b, are required. The mean for a combination can again be obtained by summing the treatment totals containing that combination, and dividing by the number of observations that have been added. For example, the mean for a_0b_0 is the total of $a_0b_0c_0$ and $a_0b_0c_1$ divided by 2×4 (i.e. 2 treatment totals each formed from 4 replications). The S.E. of such means is $\sqrt{(\frac{1}{8}\text{M.S.}_E)} = \sqrt{(\frac{1}{8} \times 0\cdot610)} = \pm 0\cdot276$ in this case. It is usual to complete the display by putting in the marginal means and their S.E.s. For the present experiment the tables would be given as:

	a_0	a_1	Mean			a_0	a_1	Mean	
b_0	9·99	10·70	10·34	±0·195	c_0	10·23	10·76	10·49	±0·195
b_1	9·35	11·25	10·30		c_1	9·11	11·19	10·15	
	(±0·276)					(±0·276)			
Mean	9·67	10·98			Mean	9·67	10·98		
	±0·195					±0·195			

Similar tables can be constructed to show a three-factor interaction, but only rarely do such interactions give easily understood biological information directly. They usually show that further experiments with more detailed null hypotheses are required.

This method of obtaining the values for display is perfectly straightforward and takes little time if there are only two or three factors, each at two levels in a simple experimental design, but it becomes unnecessarily tedious when there are many factors, and impossible with some complex designs. The method to adopt then involves building up the tables from the effects calculated as on p. 91. First the total effects are converted to mean effects. If we define the mean main effect of the factor A as the difference between the means of the two levels of that factor, it will be

$$\frac{\text{Totals of } a_1b_0c_0 + a_1b_0c_1 + a_1b_1c_0 + a_1b_1c_1}{4r} - \frac{\text{Totals of } a_0b_0c_0 + a_0b_0c_1 + a_0b_1c_0 + a_0b_1c_1}{4r}$$

where r is the replication of each combination, which is exactly our total effect of A divided by $4r$. The numerator is obvious from the construction of the contrast, where we gave the coefficient $+1$ to all combinations containing a_1 and -1 to all containing a_0. The divisor is the number of combinations of factors other than A multiplied by the number of replications or, as we saw earlier, the sum of the positive coefficients applied to individual values. Thus, for a 2^n experiment with r replications, a mean main effect on this basis is

$$\frac{\text{the total effect}}{r \times 2^{n-1}}$$

Although it is more difficult to see biological reasons for it, there is less arithmetical confusion if mean interaction effects are obtained in the same way with the same divisor.

Thus we set out an additional column to the table on p. 91 to give

FACTORIAL ARRANGEMENT OF TREATMENTS

the general mean in the first row, i.e.

$$\frac{\text{total}}{r \times 2^n} = \frac{\text{total}}{32}$$

and mean effects in the other rows by dividing each total effect by $r \times 2^{n-1} = 16$.

	Total effect	Mean effect	S.E.
Total	330·3	10·32 (M)	±0·138
A	+20·9	+1·31	
B	−0·7	−0·04	
AB	+9·5	+0·59	
C	−5·5	−0·34	±0·276
AC	+12·3	+0·77	
BC	−4·5	−0·28	
ABC	+0·1	+0·01	

S.E.s can be worked out in the usual way; for the general mean the S.E. will be $\sqrt{(\frac{1}{32}\text{M.S.}_E)}$. All the other effects are differences between two means of $r \times 2^{n-1}$ or 16 observations, so their S.E. will be $\sqrt{(\frac{2}{16}\text{M.S.}_E)}$.

Now suppose we wanted to display the means for a_0 and a_1. From the method of calculating the effect it is clear that the general mean

$$M = \tfrac{1}{2}(\text{mean for } a_1 + \text{mean for } a_0)$$

or

$$2M = \text{mean for } a_1 + \text{mean for } a_0$$

and

$$A = \text{mean for } a_1 - \text{mean for } a_0$$

adding

$$2M + A = 2 \times \text{mean for } a_1$$

and subtracting

$$2M - A = 2 \times \text{mean for } a_0$$

or

$$\text{mean for } a_1 = M + \tfrac{1}{2}A = 10\cdot32 + 0\cdot65 = 10\cdot97$$

and

$$\text{mean for } a_0 = M - \tfrac{1}{2}A = 10\cdot32 - 0\cdot65 = 9\cdot67$$

as we obtained by the other method.

Thus the overall mean for the higher level of any factor is easily obtained by adding half the mean main effect of that factor to the general mean, and the lower level by subtracting half the mean main effect from the general mean. Standard errors can also be quickly obtained for mean values by using the general rule for S.E.s of linear functions given on p. 76. If the mean for a_0 is $M+\frac{1}{2}A$ then its S.E. must be

$$\sqrt{[(\text{S.E.}_M)^2 + (\tfrac{1}{2})^2 (\text{S.E.}_A)^2]}$$

where S.E._M and S.E._A are the S.E.s for M and A respectively, i.e.

$$\sqrt{(\tfrac{1}{32}\text{M.S.}_E + \tfrac{1}{4} \times \tfrac{2}{16}\text{M.S.}_E)} = \sqrt{(\tfrac{1}{16}\text{M.S.}_E)} = 0\cdot 195$$

as calculated earlier.

If a two-factor interaction (say AB) is significant, we need means for combinations of two factors. Algebra similar to that used above will show that

mean for $a_0 b_0 = M - \tfrac{1}{2}A - \tfrac{1}{2}B + \tfrac{1}{2}AB = 10\cdot32 - 0\cdot65 + 0\cdot02 + 0\cdot30 = 9\cdot99$

mean for $a_1 b_0 = M + \tfrac{1}{2}A - \tfrac{1}{2}B - \tfrac{1}{2}AB = 10\cdot32 + 0\cdot65 + 0\cdot02 - 0\cdot30 = 10\cdot69$

mean for $a_0 b_1 = M - \tfrac{1}{2}A + \tfrac{1}{2}B - \tfrac{1}{2}AB = 10\cdot32 - 0\cdot65 - 0\cdot02 - 0\cdot30 = 9\cdot35$

mean for $a_1 b_1 = M + \tfrac{1}{2}A + \tfrac{1}{2}B + \tfrac{1}{2}AB = 10\cdot32 + 0\cdot65 - 0\cdot02 + 0\cdot30 = 11\cdot25$

The sign given to an effect here is the sign given to the combination when that effect was calculated, and can be got from a table of coefficients. The expressions are easily written down directly by putting M in every case, then of the letters involved write $-\tfrac{1}{2}$ if the lower level is present and $+\tfrac{1}{2}$ if the higher level is present, and multiply together the signs given to two main effects to obtain the sign for the interaction between these two factors, e.g. for $a_0 b_0$, a is at the lower level, so we need $-\tfrac{1}{2}A$; B is at the lower level so we need $-\tfrac{1}{2}B$; and AB must be multiplied by $(-)(-)\tfrac{1}{2}$, i.e. we need $+\tfrac{1}{2}AB$.

The S.E. of individual means in the body of such a table would be

$$\sqrt{[(\text{S.E.}_M)^2 + (\tfrac{1}{2})^2 (\text{S.E.}_A)^2 + (\tfrac{1}{2})^2 (\text{S.E.}_B)^2 + (\tfrac{1}{2})^2 (\text{S.E.}_{AB})^2]}$$
$$= \sqrt{(\tfrac{1}{32}\text{M.S.}_E + \tfrac{1}{4} \times \tfrac{2}{16}\text{M.S.}_E + \tfrac{1}{4} \times \tfrac{2}{16}\text{M.S.}_E + \tfrac{1}{4} \times \tfrac{2}{16}\text{M.S.}_E)} = \sqrt{(\tfrac{1}{8}\text{M.S.}_E)}$$

as we calculated earlier.

It was mentioned earlier that factorial arrangements might be used to test definite null hypotheses other than the simple ones that there are no main effects and interactions. For example, the null hypotheses might include the statement that a does not affect growth in the presence of b. Such comparisons can be easily calculated and tested by using the mean effects. This particular hypothesis requires a test of the difference between the mean of all observations in the combinations $a_1b_1c_0$ and $a_1b_1c_1$ on the one hand and the mean of those in $a_0b_1c_0$ and $a_0b_1c_1$ on the other. Since our mean effects are averaged over all other factors, this reduces to a comparison of the means for a_1b_1 and a_0b_1 as defined above and, in terms of effects, becomes

$$a_1b_1 - a_0b_1 = M + \tfrac{1}{2}A + \tfrac{1}{2}B + \tfrac{1}{2}AB - (M - \tfrac{1}{2}A + \tfrac{1}{2}B + \tfrac{1}{2}AB)$$
$$= A + AB = +1\cdot 31 + 0\cdot 59 = 1\cdot 90$$

and its standard error is

$$\sqrt{[(\text{S.E.}_A)^2 + (\text{S.E.}_{AB})^2]} = \sqrt{(\tfrac{2}{16}\text{M.S.}_E + \tfrac{2}{16}\text{M.S.}_E)} = \sqrt{(\tfrac{1}{4} \times 0\cdot 610)} = \pm 0\cdot 391$$

These expressions are most easily written down by first writing the contrast required, e.g. $a_1b_1 - a_0b_1$. The main effects to be included are those in which the levels differ on the two sides of the contrast, in this case $+\tfrac{1}{2}A - (-\tfrac{1}{2}A) = A$ but, since both sides have b_1, B will not be included because $+\tfrac{1}{2}B - (+\tfrac{1}{2}B) = 0$. Interactions will be included if the sign of the interaction effect differs between the two sides, in this case because a_1b_1 are both upper levels, $\tfrac{1}{2}AB$ is positive, whilst in a_0b_1 it is negative, so in the whole expression we have $(+\tfrac{1}{2}AB) - (-\tfrac{1}{2}AB) = AB$.

Likewise the effect of a with b absent $(a_1b_0 - a_0b_0)$ is $A - AB$. Taking a more complicated one, a with b and c absent (i.e. $a_1b_0c_0 - a_0b_0c_0$), a is at different levels so $(+\tfrac{1}{2}A) - (-\tfrac{1}{2}A) = A$ is included; b and c are both at the same level so are not included, AB is negative in the first and positive in the second, so we have $-AB$, likewise $-AC$; BC is positive in both so does not appear, but ABC is positive in the first $[(+1)(-1)(-1)]$ and negative in the second $[(-1)(-1)(-1)]$ so we have

$$a_1b_0c_0 - a_0b_0c_0 = A - AB - AC + ABC$$

In some more complicated experiments, where there are some factors with more than two levels, as well as two-level factors, it may be arithmetically easier to define the mean effect as the deviation of the mean of the observations on the higher level from the general mean. Then (giving the symbols primes) $A' = $ total effect $A/(r \times 2^n)$ and the mean for $a_0 = M - A'$ and mean for $a_1 = M + A'$, i.e. the halves have been already taken account of, so easing the arithmetic if many combination means have to be derived. We should use whichever system is easier for the problem in hand, but care must be taken to state clearly what the effect means in biological terms when the results are communicated to anyone else.

It can be seen that a factorial arrangement of treatments can be used for solving many sorts of problems in biology. It should be stressed that it is just as important to state the null hypotheses precisely in this sort of design as in an experiment which has a collection of treatments demanding no obvious orthogonal contrasts. Selecting parts of interactions for tests, simply because the data suggest that there may be a difference, will, like the many haphazard t-test situations, give too much weight to chance variation in drawing the conclusions. It is far better to design a new experiment in which this interaction can be examined deliberately and more fully.

CHAPTER SEVEN

USE OF ORTHOGONAL POLYNOMIALS, INTERACTIONS AND REGRESSION

There is yet another way of dividing treatment sums of squares into orthogonal components, a way which is very useful if we have hypotheses about the relationship of some measure to the level of a particular factor. Suppose, for instance, that we wanted to find the relationship between growth of a micro-organism and temperature. Our biology would tell us that many different relationships could exist but, if we know very little about the problem, it is usually best to start by considering a general polynomial relationship of the form $y = a + bx + cx^2 + dx^3 + \ldots + qx^p$, where y is the variate we are measuring and x the level of the treatment we are interested in. In our case $y =$ increase in weight of micro-organism; $x =$ the temperature, and a, b, c, etc., are constants. Looking at this geometrically, we see that if all the terms beyond a are zero then $y = a$, a constant, and there is no relationship between y and x because y has the same value whatever value x has. If, however, b has a value other than zero, but c, d and beyond are zero, $y = a + bx$ and is said to vary linearly with x, giving relations such as (a) or (b) in figure 7.1, where y increases or decreases by b for every unit increase in x, respectively.

At the next stage, both b and c might have values other than zero, whilst d and beyond are all zero; then $y = a + bx + cx^2$ and possible curves are (c), (d), (e) and (f) of figure 7.1; the expression is then said to be *quadratic*.

If d also has a non-zero value, we have a cubic expression and cubic curve such as is shown in figure 7.1 (g), but variously modified according to the signs and relative values of a, b, c and d. From there on every additional term puts another bend in the curve, but in real life we are seldom interested in anything beyond quadratic. So in our example we might make three null hypotheses: (1) there is no linear relationship between growth and temperature; (2) there is no quadratic effect; (3) there are no deviations from the linear and quadratic effects. Now we know that we can fit a straight line to two points exactly, and fit a quadratic curve to three points exactly, so if we want to see if there are any deviations

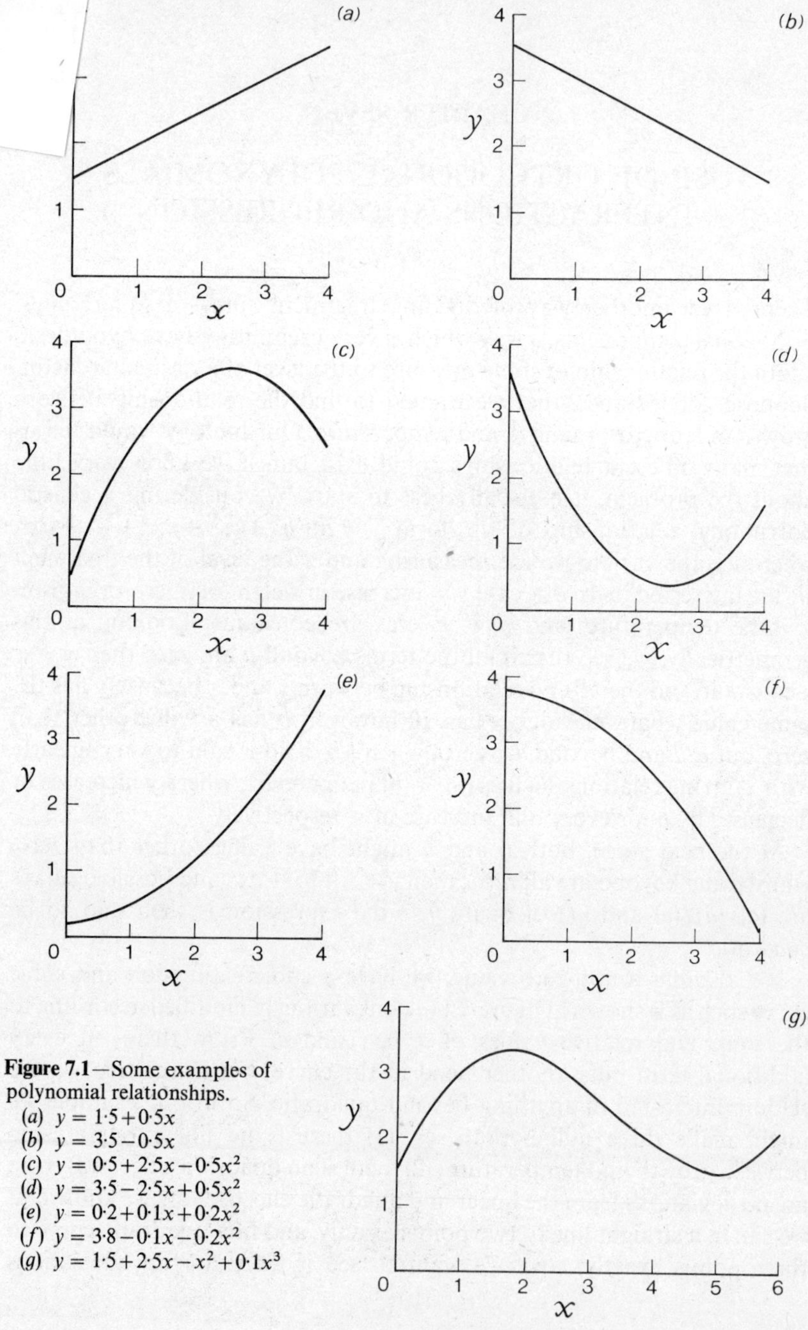

Figure 7.1—Some examples of polynomial relationships.
(a) $y = 1.5 + 0.5x$
(b) $y = 3.5 - 0.5x$
(c) $y = 0.5 + 2.5x - 0.5x^2$
(d) $y = 3.5 - 2.5x + 0.5x^2$
(e) $y = 0.2 + 0.1x + 0.2x^2$
(f) $y = 3.8 - 0.1x - 0.2x^2$
(g) $y = 1.5 + 2.5x - x^2 + 0.1x^3$

from the linear and quadratic we must have at least four points and, of course, we must have the necessary replication of all treatments to give an estimate of natural variation with which to do the tests. So suppose we choose four equally spaced levels of temperature: 10°, 20°, 30° and 40°C, and have six replications. Now we have a basic analysis of variance of

Source of variation	D.F.
Treatments	3
Error	20

but must find a way of obtaining sums of squares due to the linear effect, due to the quadratic effect, and due to deviations from the linear and quadratic, which in our case is simply those due to the cubic effect, since the four points must fit a cubic perfectly. We cannot just solve for a, b, c, etc., directly because we need three orthogonal contrasts so that we can test each separately. We therefore find an equivalent line $Y = A + B\xi_1 + C\xi_2 + D\xi_3$, where ξ_1 is a linear function of x, ξ_2 a quadratic function, and so on, and the ξs are orthogonal to each other. To obtain the ξs we first write down the four temperatures in order (Table 7.1, column 1). Then to ease the arithmetic we can bring the range of temperatures down to the smallest possible integers by dividing by 10 (column 2) and then subtracting 1 from each (column 3) (the usual processes of coding).

Now any linear effect must be of the form $\xi_1 = a + bx$, where a and b are constants, and we could divide by the constant b without altering the relationship; so we can simplify to $\xi_1 = a + x$, x being the level of temperature. Substituting the numerical values for x we get column (4). We now remember that the first rule of orthogonal contrasts was that the coefficients of each contrast must sum to zero. Therefore $4a + 6 = 0$ and $a = -\frac{6}{4} = -\frac{3}{2}$. Now substitute for a in each row and produce the next column. We can get rid of the fractions by multiplying by 2, and arrive finally at the coefficients for the linear contrast of $-3, -1, +1, +3$. This set is usually referred to as ξ'_1. ξ_1 is taken to mean the basic coefficient, in our case $-\frac{3}{2}, -\frac{1}{2}, +\frac{1}{2}, +\frac{3}{2}$, and, when common factors are removed, the coefficients are given the symbol ξ'; the multiplier used to do this is referred to as λ (in this case 2). These λs are necessary to enable simple methods for getting values for intermediate levels of x to be used.

Proceeding to the quadratic contrast, it must be of the form $\xi_2 = a + bx + x^2$ (again dividing through by the last constant to make the arithmetic easier). We substitute our numerical values for x and obtain column (7), which again must sum to zero. But now we have two

Table 7.1—Derivat

(1) Temperature	(2) x	(3)	(4) $a+x$	(5) ξ_1	(6) ξ_1'	(7) $a+bx+x^2$	(8) $\xi_1'(a+bx+x^2)$	(9) ξ_2'
10	1	0	a	$-\frac{3}{2}$	-3	a	$-3a$	$+1$
20	2	1	$a+1$	$-\frac{1}{2}$	-1	$a+b+1$	$-a-b-1$	-1
30	3	2	$a+2$	$+\frac{1}{2}$	$+1$	$a+2b+4$	$+a+2b+4$	-1
40	4	3	$a+3$	$+\frac{3}{2}$	$+3$	$a+3b+9$	$+3a+9b+27$	$+1$
Total	—	—	$4a+6$	0	0	$4a+6b+14$	$10b+30$	0
					$\lambda = 2$			$\lambda = 1$

unknowns, a and b, so obviously cannot solve for these with a single equation. But we remember the second rule of orthogonal contrasts, that the product of the coefficients of any two contrasts must sum to zero, so multiply the expressions for the quadratic coefficients by the coefficients we obtained for the linear contrast (ξ_1' in column 6) to give column (8). Adding these we get $10b + 30 = 0$, so $b = -3$, and substituting in the previous equation $4a - 18 + 14 = 0$, so $a = 1$. Now substitute these values for a and b in our original equations in column (7) and we get ξ_2' straight away as $+1, -1, -1, +1$, since there are no common factors to remove (column 9).

Similarly, to get the cubic coefficients we take $\xi_3 = a + bx + cx^2 + x^3$, and obtain column (10). Multiply by ξ_1' to obtain column (11) and by ξ_2' to obtain column (12).

Then from column (12), $4c + 18 = 0$, so $c = -\frac{9}{2}$.

Substituting in the total of column (11) gives

$$10b - \frac{30 \times 9}{2} + 88 = 0 \quad \text{or} \quad b = \frac{47}{10}$$

and from the total of column (10)

$$4a + \frac{6 \times 47}{10} - \frac{14 \times 9}{2} + 36 = 0 \quad \text{or} \quad a = -\frac{3}{10}$$

Substituting for a, b and c in the original equations of column (10) gives column (13).

$$\begin{aligned}
a &= -\tfrac{3}{10} \\
a+b+c+1 &= \tfrac{1}{10}(-3+47-45+10) &= +\tfrac{9}{10} \\
a+2b+4c+8 &= \tfrac{1}{10}(-3+94-180+80) &= -\tfrac{9}{10} \\
a+3b+9c+27 &= \tfrac{1}{10}(-3+141-405+270) &= +\tfrac{3}{10}
\end{aligned}$$

USE OF ORTHOGONAL POLYNOMIALS

(10)	(11)	(12)	(13)	(14)
$+bx+cx^2+x^3$	$\xi_1'(a+bx+cx^2+x^3)$	$\xi_2'(a+bx+cx^2+x^3)$	ξ_3	ξ_3'
a	$-3a$	$+a$	$-\frac{3}{10}$	-1
$a+b+c+1$	$-a-b-c-1$	$-a-b-c-1$	$+\frac{9}{10}$	$+3$
$a+2b+4c+8$	$+a+2b+4c+8$	$-a-2b-4c-8$	$-\frac{9}{10}$	-3
$a+3b+9c+27$	$+3a+9b+27c+81$	$+a+3b+9c+27$	$+\frac{3}{10}$	$+1$
$a+6b+14c+36$	$10b+30c+88$	$4c+18$	0	0
				$\lambda = \frac{10}{3}$

We can divide through by $\frac{3}{10}$ to give us our ξ_3' values in column (14). This method can be used no matter what the spacing of the levels is, e.g. we could get suitable orthogonal polynomials for temperatures of 10, 20, 50, 70°C, but it is rather laborious, and for evenly spaced levels all the necessary orthogonal polynomials have been worked out and published in Fisher and Yates' Tables (1963).

Returning to our problem, we should set out the treatment totals and the orthogonal polynomial coefficients as in Table 7.2.

Table 7.2—Treatment totals (y) and coefficients (ξ')

	Treatments					
	10°C	20°C	30°C	40°C	$\Sigma \xi' y$	$\frac{(\Sigma \xi' y)^2}{r \Sigma \xi'^2} = $ S.S.
y	307	391	406	403		
ξ_1'	-3	-1	$+1$	$+3$	$+303$	$\frac{(+303)^2}{6 \times 20} = 765$
ξ_2'	$+1$	-1	-1	$+1$	-87	$\frac{(-87)^2}{6 \times 4} = 315$
ξ_3'	-1	$+3$	-3	$+1$	$+51$	$\frac{(+51)^2}{6 \times 20} = 22$

Then the treatment totals are multiplied by the sets of coefficients in turn to obtain the total effects. Thus the linear effect is

$$\Sigma \xi_1' y = -3 \times 307 - 1 \times 391 + 1 \times 406 + 3 \times 403 = +303$$

The sum of squares for each of these effects is simply the effect squared and divided by the sum of the squares of the coefficients applied to individual values, as usual, i.e. $(\sum \xi' y)^2 / r \sum \xi'^2$, where r is the number of replications. For the linear effect the S.S. is therefore

$$\frac{(+303)^2}{6 \times 20} = 765$$

These sums of squares can be entered into an analysis of variance table (Table 7.3), the total sum of squares worked out in the usual way, the error S.S. obtained by difference, and the effects tested by means of F-tests.

Alternatively, sum of squares for treatments as a whole can be calculated from the treatment totals, the error S.S. obtained by the difference between this value and the total S.S., and the effects tested by t-tests. The S.E. of a total effect, following the usual rule for linear contrasts will obviously be $\sqrt{(\text{M.S.}_E \times r \times \sum \xi'^2)}$.

Table 7.3—Analysis of variance

Source of variation	D.F.	S.S.	M.S.	F
Linear effect	1	765		14·54**
Quadratic effect	1	315		5·99*
Cubic effect	1	22		0·42
Error	20	1052	52·6	
Total	23	2154		

Indeed, there may be great economy in a situation involving many levels by obtaining the overall sum of squares for treatments and the error S.S. before working out the contrasts, even if the F-test is to be used, because it is then possible to see after each effect has been calculated if there are sufficient sums of squares available to enable any higher-order effect to be significant; if not there may be no point in working out the contrast. In the present example, after working out the quadratic we should have seen that only 22 of the treatment S.S. remained, so, knowing that the error M.S. was 52·6, we could see immediately that there could be no more significant effects since, when $n_1 = 1$, F is never less than 3·84 at the 5 per cent level. We need not waste our time calculating the cubic effect.

Completing Table 7.3 shows F-values greater than the 1 per cent and 5 per cent values for the linear and quadratic effects respectively, so we can say that assuming a polynomial relationship between growth of the microorganism and temperature, if there is not truly a linear and quadratic term in the relationship, we have witnessed a very unlikely event. On the other hand, the very small F-value for the cubic effect confirms our

USE OF ORTHOGONAL POLYNOMIALS 107

belief in the third null hypothesis, namely, that there are no deviations, other than the natural variation, from a relation of the form $y = a+bx+cx^2$.

The results of this experiment could be shown graphically by plotting the curve of best fit. Our orthogonal polynomials again make this very simple. We can fit the curve $Y = A + B' \xi'_1 + C' \xi'_2$ directly. We include all the ξ's up to the highest order considered significant, and first work out the mean effects.

The general mean (A)

$$= \frac{\sum y}{n} = \frac{1507}{24} = 62.8$$

For fitting the curve the mean polynomial effects are best defined as the total effect divided by the sum of squares of the coefficients, so the mean linear effect (B')

$$= \frac{\sum \xi'_1 y}{r \sum \xi'^2_1} = \frac{+303}{6 \times 20} = +2.5$$

and the mean quadratic effect (C')

$$= \frac{\sum \xi'_2 y}{r \sum \xi'^2_2} = \frac{-87}{6 \times 4} = -3.6$$

We can then write out a table giving the mean effects and their appropriate ξ' coefficients thus:

	Temperature	A	B'	C'	
		+62·8	+2·5	−3·6	
Y_1	10°	+1	−3	+1	51·7
Y_2	20°	+1	−1	−1	63·9
Y_3	30°	+1	+1	−1	68·9
Y_4	40°	+1	+3	+1	66·7

Multiplying A, B' and C' by the appropriate orthogonal polynomial coefficients, we get values which enable us to draw the curve of best fit (figure 7.2). We can put dots for the actual values, and can see at once that the deviations from this curve are small, as our analysis of variance indicated. This is often referred to as a *response curve*, and we include all the orthogonal polynomials up to and including the highest degree that achieves significance. The S.E. for any point on the curve is

$$\sqrt{\text{M.S.}_E \left(\frac{1}{n} + \frac{\xi'^2_1}{r \sum \xi'^2_1} + \frac{\xi'^2_2}{r \sum \xi'^2_2} + \cdots \right)}$$

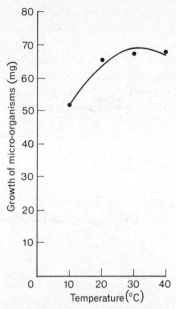

Figure 7.2—Effect of temperature on growth of a micro-organism: ——— line of best fit; ● observed values.

where n is the total number of observations, i.e. the number on which A, the general mean, is based. When we have plotted our curves, we have a good knowledge of what happens within the limits set by the treatments used. (It is very dangerous to extrapolate the curve outside these limits.) It is possible to convert these functions of ξ' back to functions of x, but such a conversion is seldom of use practically; the drawn curve gives all the information available.

Interactions involving orthogonal polynomials

There is an armoury of orthogonal contrasts that a biologist can use to answer direct questions, all of the type which can be answered "yes" or "no", or for providing estimates of direct effects. It is also possible to combine factors and draw out orthogonal contrasts from the interactions. Thus, suppose that in the experiment we have just been discussing, we wished to find the relationship of temperature to growth of two different species; we then have an interaction problem and the null hypotheses could be:

(1) The organisms on average show no linear effect, no quadratic effect, and no deviations from linear and quadratic, as before.

USE OF ORTHOGONAL POLYNOMIALS   109

(2) The linear effect is the same for the two species.
(3) The quadratic effect is the same for the two species.
(4) The deviations from the linear and quadratic are the same for the two species.

The break-up for the degrees of freedom for treatments will be:

	D.F.
Species (S)	1
Temperature linear (T')	1
Temperature quadratic (T'')	1
Temperature cubic (T''')	1
$S \times T'$	1
$S \times T''$	1
$S \times T'''$	1
Total treatments	7

To obtain the effects we need a table of coefficients:

	$s_0 t_0$	$s_0 t_1$	$s_0 t_2$	$s_0 t_3$	$s_1 t_0$	$s_1 t_1$	$s_1 t_2$	$s_1 t_3$
S	−1	−1	−1	−1	+1	+1	+1	+1
T'	−3	−1	+1	+3	−3	−1	+1	+3
T''	+1	−1	−1	+1	+1	−1	−1	+1
T'''	−1	+3	−3	+1	−1	+3	−3	+1
ST'	+3	+1	−1	−3	−3	−1	+1	+3
ST''	−1	+1	+1	−1	+1	−1	−1	+1
ST'''	+1	−3	+3	−1	−1	+3	−3	+1

This is built up in the same way as the table of coefficients for the 2^n design. For S, a two-level factor, s_1 is given the coefficient $+1$ and s_0 is given -1. The T-contrasts have the coefficients for linear, quadratic and cubic effects obtained from tables of orthogonal polynomials (such as Fisher and Yates, 1963) if the temperatures are spaced in equal steps as in the present example, or calculated as shown on pp. 104–5 if unequal steps have been used. The coefficients for interactions are obtained simply by multiplying together those for the two main effects involved, i.e. ST' has $(-1)(-3), (-1)(-1), (-1)(+1)\ldots$.

We could set out the appropriate treatment total at the head of each column, multiply each total by its coefficient, and add in the usual way. However, whenever there is an interaction between two factors, one of which has only two levels, there is a much simpler method of calculation. If we divide the table between $s_0 t_3$ and $s_1 t_0$, the numbers on the left are always repeated on the right, for main effects of T their signs are the same, whereas for interaction effects their signs are reversed. When they have the same signs, we could get the effect by adding $s_0 t_0 + s_1 t_0$;

$s_0t_1+s_1t_1$; $s_0t_2+s_1t_2$; and $s_0t_3+s_1t_3$, and then multiplying each sum by its appropriate coefficient. Similarly, when they have opposite signs, we could get the effects by taking the differences $s_1t_0-s_0t_0$; $s_1t_1-s_0t_1$; $s_1t_2-s_0t_2$; $s_1t_3-s_0t_3$, and applying the coefficients of the right-hand side of the table. This saves quite a bit of arithmetic. In our example we should set out the treatment totals in a two-way table:

	t_0	t_1	t_2	t_3	Total
s_0	307	391	406	403	1507
s_1	261	388	431	429	1509
Total	568	779	837	832	3016
Difference	−46	−3	+25	+26	+2
ξ'_1	−3	−1	+1	+3	
ξ'_2	+1	−1	−1	+1	
ξ'_3	−1	+3	−3	+1	

Obtaining the Grand Total and Grand Difference in two ways gives a check to the arithmetic. We now apply the ξ'-values to the sums and differences in turn. Thus, given six replications of each treatment combination:

using the totals

$$\text{effect of } T' = -3 \times 568 - 1 \times 779 + 1 \times 837 + 3 \times 832 = +850$$

with sum of squares

$$\frac{850^2}{12 \times 20} = 3010 \cdot 42$$

$$\text{effect of } T'' = +1 \times 568 - 1 \times 779 - 1 \times 837 + 1 \times 832 = -216$$

$$\text{S.S.} = \frac{(-216)^2}{12 \times 4} = 972 \cdot 00$$

$$\text{effect of } T''' = -1 \times 568 + 3 \times 779 - 3 \times 837 + 1 \times 832 = +90$$

$$\text{S.S.} = \frac{90^2}{12 \times 20} = 33 \cdot 75$$

using the differences

$$\text{effect of } ST' = -3 \times (-46) - 1 \times (-3) + 1 \times (+25) + 3 \times (+26) = +244$$

$$\text{S.S.} = \frac{(+244)^2}{12 \times 20} = 248 \cdot 07$$

$$\text{effect of } ST'' = +1 \times (-46) - 1 \times (-3) - 1 \times (+25) + 1 \times (+26) = -42$$

$$\text{S.S.} = \frac{(-42)^2}{12 \times 4} = 36 \cdot 75$$

effect of $ST''' = -1\times(-46)+3\times(-3)-3\times(+25)+1\times(+26) =$

$$\text{S.S.} = \frac{(-12)^2}{12\times 20} = 0\cdot 60$$

and finally the total effect of $S = +2$ with S.S. of

$$\frac{(+2)^2}{2\times 24} = 0\cdot 08$$

Each of these sums of squares has 1 degree of freedom, so the mean square is the same, and they are orthogonal. All can be tested against the error mean square, which was 52·57 in this experiment, with an F-test, when we find that $T'(F = 57\cdot 3)$, $T''(F = 18\cdot 5)$ and $ST'(F = 4\cdot 7)$ are all significant. We can claim true linear and quadratic effects on average for the organisms, but also know that the extent of the linear effect is not the same for the two species. Since an interaction is significant, it will require two curves of best fit to demonstrate the results. It was stated earlier that when fitting a single curve, we include in the model all orthogonal polynomials up to and including the highest order that is significant (i.e. judged to be a true effect, using whatever level of probability the experimenter is satisfied with), irrespective of whether the lower-order ones are significant or not. Likewise in an interaction situation we include all polynomials up to the highest found significant and include the appropriate main effects of the factors making up the significant interactions. Thus had ST'' been significant, in addition to ST'' we should include ST', S, T'' and T', irrespective of whether any or all of them were significant or not.

In the present case ST' is the highest-order interaction effect which is significant, so the model must include ST', S and T'; but T'' is itself significant, so must be included, and the model becomes

$$Y = \text{general mean} + k_1 S + k_2 T' + k_3 T'' + k_4 ST'$$

where the ks are the coefficients in the table on p. 109 and S, T', T'' and ST' are the mean effects calculated as the total effects divided by $r\sum k^2$. Thus

$$S = \frac{+2}{24\times 2} = +0\cdot 04; \quad T' = \frac{+850}{12\times 20} = +3\cdot 54$$

$$T'' = \frac{-216}{12\times 4} = -4\cdot 50; \quad \text{and} \quad ST' = \frac{+244}{12\times 20} = +1\cdot 02$$

and these, together with the mean, can be set out at the head of a table and the points through which the curves should pass calculated in the same way we used for the single-species case.

	General Mean	S	T'	T"	ST'	Y
	62·83	+0·04	+3·54	−4·50	+1·02	
$s_0 t_0$	+1	−1	−3	+1	+3	50·7
$s_0 t_1$	+1	−1	−1	−1	+1	64·8
$s_0 t_2$	+1	−1	+1	−1	−1	69·8
$s_0 t_3$	+1	−1	+3	+1	−3	65·8
$s_1 t_0$	+1	+1	−3	+1	−3	44·7
$s_1 t_1$	+1	+1	−1	−1	−1	62·8
$s_1 t_2$	+1	+1	+1	−1	+1	71·9
$s_1 t_3$	+1	+1	+3	+1	+3	72·0

Figure 7.3 shows the curves. It can be seen that s_1 had a greater linear response to temperature than s_0, because it grew less at the low temperature but more at the high. The statistical analysis has given us confidence in this opinion for the species in general, because we know that, if the two species do respond to temperature in the same way, we have witnessed an event which could occur less than once in 20 times due to chance.

When we have two or more factors all at more than two levels, we

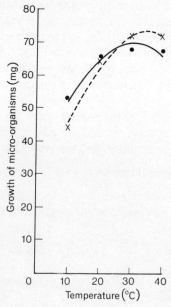

Figure 7.3—Effect of temperature on growth of two species of micro-organism: ——— line of best fit for s_0; ● observed values; - - - - - line of best fit for s_1; × observed values.

USE OF ORTHOGONAL POLYNOMIALS

can still calculate interaction effects. If the factors themselves can be broken down to orthogonal contrasts, each with one degree of freedom, then the interactions can be broken down likewise. Suppose we had two factors (A and B) each at three levels, and we decided to split the sum of squares for each factor into components for linear and quadratic effects. Then we could have a complete analysis of variance of the treatments:

	D.F.
A linear (A')	1
A quadratic (A'')	1
B linear (B')	1
B quadratic (B'')	1
A linear × B linear ($A'B'$)	1
A linear × B quadratic ($A'B''$)	1
A quadratic × B linear ($A''B'$)	1
A quadratic × B quadratic ($A''B''$)	1
Total treatment	8

The method of calculation would be to use a table of coefficients as in the 2^n design. For three equally spaced levels we would find in Fisher and Yates' tables that the linear has the coefficients $-1, 0, +1$ while for quadratic they are $+1, -2, +1$, so we could set out the coefficients for our 9 treatment totals as:

	a_0b_0	a_0b_1	a_0b_2	a_1b_0	a_1b_1	a_1b_2	a_2b_0	a_2b_1	a_2b_2	$\sum k^2$
A'	−1	−1	−1	0	0	0	+1	+1	+1	6
A''	+1	+1	+1	−2	−2	−2	+1	+1	+1	18
B'	−1	0	+1	−1	0	+1	−1	0	+1	6
B''	+1	−2	+1	+1	−2	+1	+1	−2	+1	18
$A'B'$	+1	0	−1	0	0	0	−1	0	+1	4
$A'B''$	−1	+2	−1	0	0	0	+1	−2	+1	12
$A''B'$	−1	0	+1	+2	0	−2	−1	0	+1	12
$A''B''$	+1	−2	+1	−2	+4	−2	+1	−2	+1	36

Having multiplied the treatment totals by these coefficients, we obtain the total effect in each case. Dividing this total effect by the number of replications times the sum of the squares of the coefficients ($r\sum k^2$) gives a mean effect, and we get the sum of squares for each effect by squaring the total effect and dividing by $r\sum k^2$.

When we interpret the results, significance for any main effect (A', A'', B' or B'') will indicate that there is a real effect on average over the three levels of the other factor. Significance of an interaction will show that the effect shown by one factor differs from level to level of the other factor, e.g. if A' and $A'B'$ were both significant and positive we might have figure 7.4(a) or, in words, the increase in yield per unit increase in A

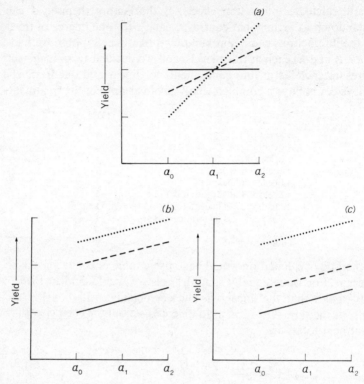

Figure 7.4—Some examples of the effects of interactions.
(a) When A' and $A'B'$ are both significant and positive.
(b) When A' and B' are significant and positive, and $A'B''$ is significant and negative.
(c) When A', B' and $A'B''$ are all significant and positive.
——— effect of A at b_0; ----- effect of A at b_1; ········ effect of A at b_2.

becomes greater as the level of B is increased. If A' and B' were significant and positive, and $A'B''$ was significant, we might have figure 7.4(b) or (c).

Some other sorts of interaction

Occasionally a biologist has a problem which involves two or more factors, one of which is not susceptible to being broken down to individual contrasts, e.g. a geneticist may wish to know if all his supposedly different genotypes (G) give the same response to two nutrients (A and B), and at the same time it would be wise to test if the AB-interaction was also the same for each genotype.

INTERACTIONS

Supposing he had four genotypes, he would require 16 treatments made up of all combinations of the 4 genotypes (g_0, g_1, g_2 and g_3) × 2 levels of A (a_0 and a_1) × 2 levels of B (b_0 and b_1). Three replications of these in a completely randomized design giving 32 D.F. for error might be adequate. We would set out the treatment totals in a two-way table such as Table 7.4.

Table 7.4—Treatment totals

	g_0	g_1	g_2	g_3	Total	A	B	AB
							Coefficients	
a_0b_0	106·0	123·5	130·6	120·0	480·1	−1	−1	+1
a_1b_0	116·9	122·4	144·1	127·8	511·2	+1	−1	−1
a_0b_1	140·2	123·4	161·3	141·8	566·7	−1	+1	−1
a_1b_1	142·3	114·6	161·3	147·9	566·1	+1	+1	+1
Total	505·4	483·9	597·3	537·5	2124·1			

The 15 D.F. for the treatments could be divided as in Table 7.6. The main effects of G obviously have 3 D.F., because there are four genotypes, and the combinations of A and B, averaged over all genotypes can be divided in the usual 2^n fashion to give main effects A and B and the interaction AB. Now we are left with the interactions between the genotypes and the nutrients which should provide the tests of the null hypotheses that the experiment was set up to test, namely, the genotypes do not differ in response to A, B or AB.

If we calculated the response to A, say, for each genotype separately, we could see if these responses varied around the mean response more than would be expected from the natural variation by making an F-test in exactly the same way as we would in a single-factor situation when the null hypothesis is that all levels of that factor come from the same population. Here we are saying all responses to A come from the same population of responses. Since there are four responses involved, the variation between them will have 3 D.F., and this conforms with our previous rule that the number of degrees of freedom for an interaction is the product of the degrees of freedom for the contrasts which make up the interaction. (G has 3 D.F., A has 1 D.F., so GA has $3 \times 1 = 3$ D.F.) Likewise GB and GAB will both have 3 D.F. and can be tested by an F-test. We see that the degrees of freedom for the individual items sum to 15, the total D.F. for treatments, suggesting that all the sums of squares have been accounted for.

The fact that the D.F. sum to the right total does not guarantee orthogonality of the split. If in doubt, it can be checked by allocating known orthogonal contrasts to the factor which has not been broken down to individual degrees of freedom, say $-1, +1, -1, +1; -1, -1, +1, +1$; and $+1, -1, -1, +1$ to g_0, g_1, g_2, g_3 in this case, and test that all $\sum k_i k_j = 0 (i \neq j)$.

To obtain the sums of squares it is best first to set out the effects in a two-way table such as Table 7.5. To obtain the effect A for g_0, we simply apply the A-coefficients $(-1, +1, -1, +1)$ to the totals for $a_0 b_0$, $a_1 b_0, a_0 b_1, a_1 b_1$, respectively in the column applying to g_0 in Table 7.4, i.e. A for $g_0 = -106 \cdot 0 + 116 \cdot 9 - 140 \cdot 2 + 142 \cdot 3 = +13 \cdot 0$. All the others are got in the same way by applying the appropriate A, B or AB-coefficients to each set of genotype totals in turn to obtain Table 7.5.

Table 7.5—Table of effects

	g_0	g_1	g_2	g_3	Total
A	+13·0	−9·9	+13·5	+13·9	+30·5
B	+59·6	−7·9	+47·9	+41·9	+141·5
AB	−8·8	−7·7	−13·5	−1·7	−31·7

The values in the total column of this table should be computed directly by applying the coefficients to the total column of Table 7.4 and checked by summing effects over all gs in the row.

Now S.S. for G will be the sum of squared g totals in Table 7.4, divided by the number of individual values which have been used to obtain each total, minus the correction factor, in exactly the way of the primrose example of Chapter 5, i.e.

$$\frac{505 \cdot 4^2 + 483 \cdot 9^2 + 597 \cdot 3^2 + 537 \cdot 5^2}{3 \times 4} - \frac{2124 \cdot 1^2}{48} = 609 \cdot 31$$

The S.S. for A is calculated as

$$\frac{(\text{total effect } A)^2}{r \sum k^2} = \frac{(+30 \cdot 5)^2}{3 \times 4 \times 4} = 19 \cdot 38$$

in the same way as we have used previously for orthogonal contrasts.

The S.S. for B and AB are calculated similarly. It should be noticed that, using these formulae, the number of replications r will be the number of replications making up the total to which the coefficients k have been applied. Since there were four gs receiving $a_0 b_0$, say, in each of three complete replicates, the $a_0 b_0$ total came from 3×4 or 12 individual observations, in the same way as the divisor for $\sum g^2$ above was 3×4 because each genotype was present with each of the four combinations of A and B in each of the three replicates. Indeed, with more complex experiments, it is safer always to consider the coefficients to be applied to individual values, since this leads to more easily remembered universal rules. Here for the S.S. for A we should imagine that every $a_0 b_0$ value was multiplied by -1 and would see immediately that there are 12 such values. Their contribution to the divisor is $12(-1)^2 = 12$, likewise $a_1 b_0$ contributes $12(+1)^2$, $a_0 b_1$ $12(-1)^2$, and $a_1 b_1$ $12(+1)^2$, making 48 in all as the divisor. In terms of applying coefficients to individuals, we have a completely universal rule that if an S.S. is to be calculated by squaring a single quantity Q, where Q is made up as a linear function of individual values, i.e.

$$Q = k_1 x_1 + k_2 x_2 + k_3 x_3 \ldots k_n x_n$$

then

$$\text{S.S.}_Q = \frac{Q^2}{\sum k^2}$$

Furthermore, if an S.S. is obtained by squaring a number of quantities, each made up of a linear function of individual values which are quite distinct, i.e. no individual value has been used to form more than one of them, again the divisor for each is the sum of squares of the coefficients, i.e.

$$\frac{Q_1^2}{\sum k_1^2} + \frac{Q_2^2}{\sum k_2^2} + \ldots + \frac{Q_n^2}{\sum k_n^2} - \frac{(\sum Q)^2}{\sum k^2}$$

In the case of S.S. for G above, each quantity squared was a total of 12 individual values, i.e.

$$(+1)x_1 + (+1)x_2 + (+1)x_3 \ldots + (+1)x_{12}$$

so the divisor was $(+1)^2 + (+1)^2 + (+1)^2 + \ldots$ up to 12 terms, which is 12.

Now we can apply this rule to obtain the S.S. for GA, the variation of A effects on the various genotypes. It will be

$$\frac{(+13 \cdot 0)^2 + (-9 \cdot 9)^2 + (+13 \cdot 5)^2 + (+13 \cdot 9)^2}{12} - \frac{(+30 \cdot 5)^2}{48} = 53 \cdot 54 - 19 \cdot 38 = 34 \cdot 16$$

The first divisor is 12 because to obtain the A-effect at g_0, say, we multiplied three $a_0 b_0$s by -1, three $a_1 b_0$s by $+1$, three $a_0 b_1$s by -1, and three $a_1 b_1$s by $+1$ and $3(-1)^2 + 3(+1)^2 + 3(-1)^2 + 3(+1)^2 = 12$. Likewise for A at g_1, g_2 and g_3. The correction factor is the S.S. for A, which we have calculated already, and this perhaps shows more clearly what we have done.

The four effects of A, one for each genotype, form the total information we have about the A-effect. We have divided it into two parts: (1) the average effect obtained from the total or mean of these four, and (2) the variation of the individual genotype values around that mean. If we adjudged this variation to be unduly large, we should not be prepared to believe that the effect of A was the same for all genotypes. S.S. for GB and GAB are worked out similarly. A check on the arithmetic can be made by calculating the S.S. for treatments from the original treatment totals and confirming that this is the same as the sum of the component parts.

The error S.S. would be calculated in the usual way; all mean squares are then calculated and F-tests carried out. In the event, GB gives an F-value larger than the value in the table for $P \leqslant 0\cdot01$, so we should find it hard to go on believing that the genotypes all respond to nutrient B in the same way. But we have no reason to reject our belief that their response to A and AB are the same, though, as A itself had no effect on average, we should beware of saying that we had added much evidence to this belief, since if A is a nutrient which usually affects the growth of this species, we would suspect that the levels we chose were too similar, or that some other limiting factor may have been present which prevented any of the genotypes from showing responses.

Table 7.6—Analysis of variance

Source of variation	D.F.	S.S.	M.S.	F
G	3	609·31	203·10	15·52***
A	1	19·38	19·38	1·48
B	1	417·13	417·13	31·87***
AB	1	20·94	20·94	1·60
GA	3	34·16	11·39	<1
GB	3	221·59	73·86	5·64**
GAB	3	5·89	1·96	<1
Total treatments	15	1328·40		
Error	32		13·09	

** $0\cdot001 < P < 0\cdot01$; *** $P < 0\cdot001$

It is accepted almost as an axiom of scientific investigation that the most clear-cut information is that which can be tested on a "true" or "false" basis. It follows that when designing experiments we can best achieve this by setting null hypotheses which result in testing the variation appropriate to single degrees of freedom. If several null hypotheses are to be tested in the same experiment, then arranging the treatment combinations so that $p-1$ useful orthogonal contrasts can be got from p treatment combinations usually leads to the most economical use of time and material. There may, however, be occasions when this is not possible. One such occasion occurs when one factor of a multifactor design needs investigation on the basis of comparing one level (often the nil level) with each of the others. Then the interaction should be dealt with likewise, using appropriate t-values to compare these levels within each combination of levels of the other factors.

Another occasion arises when both the factors of a two-factor design represent samples (possibly random samples) of two populations. For example, we might be selecting for a new strain of perennial ryegrass and at the first selection have 100 plants which are then propagated vegetatively. Before selecting further, we might ask, "Does it matter where and how they are grown for the selection experiments?" If the highest yielder in one environment is also the highest yielder in all other environments, then it would not matter where they were grown if the objective was to pick that one out. Rather than take a chance on this being so, the breeder might prefer to test if there was an interaction between clones and environment by making an experiment with perhaps 10 of the clones chosen at random as one factor and five environments, chosen nearly enough random as the other, when his important null hypothesis would be that there is no interaction between clones and environment. In such a case the interaction sum of squares has to be obtained by calculating the total S.S. for treatments (from the 50 treatment totals), the S.S. for clones from the 10 clone totals, and S.S. for environments from the five environment totals. The S.S. for the interaction = S.S. for total treatments − S.S. for clones − S.S. for environments. The interaction has $(10-1)(5-1) = 36$ D.F. and would be tested by an F-test in the usual way.

Regression

When studying the relationship between two measures, it is not always possible to get exact replicates of the treatments, as we could with the example of the effect of temperature on growth of a micro-organism. For example, in a study of rhizobium, we might be faced with the very tedious and time-consuming task of counting cells, and wish to devise a simple method based on passing light through a suspension of cells and measuring the amount of light absorbed. It might be expected that the number of cells per unit volume of suspension should vary linearly with the absorptiometer reading. It would not be possible, without a tremendous amount of work, to set up, say, four suspensions with an absorptiometer reading of 0·2, four with 0·4, four with 0·6 and four with 0·8 exactly, but a series covering the range in which we are interested could be set up and the number of cells in each estimated by the usual counting method. These same suspensions could then be measured for light absorption. We proposed a linear relationship of

$$y = a + bx$$

where y is the number of cells per ml, x is the absorptiometer reading, and a and b are constants. The statistical procedures are much the same as those used in analysis of variance with orthogonal contrasts, but the operation is usually referred to as *regression analysis*. To assess the usefulness of our method of estimating number of cells, we should need to satisfy ourselves that b, which is called the *regression coefficient*, was not zero, for y would then equal a for all suspensions. We should wish to calculate confidence limits for y so that, when we used the method, we could say within what limits the true number of cells of any suspension we measured on the absorptiometer was likely to lie. Some results of such an experiment are given in Table 7.6.

Table 7.6—Absorptiometer readings (x) and cell counts (y, 10^8/ml) for a series of suspensions

x	y	x	y	x	y
0·37	8·2	0·64	12·1	0·84	15·8
0·59	10·6	0·77	14·2	0·71	18·2
0·48	7·3	0·78	16·1	1·02	16·8
0·62	13·3	0·93	15·0	0·91	19·1
0·74	11·4	0·81	16·9	0·94	23·4
0·71	12·9				

The analysis follows from the method we used in deriving the linear effect from orthogonal polynomials (p. 107). There we used a model of $y = \bar{y} + B'\xi'_1$ and the linear effect of y on ξ'_1, i.e. B' of this expression, was

$$\frac{\sum \xi'_1 y}{r \sum \xi'^2_1}$$

when the ys were totals of r observations. This is the same as writing

$$\frac{\sum \xi'_1 y}{\sum \xi'^2_1}$$

when the ys are single observations and ξ'_1s are applied to each individually. Using the latter definition of y, the S.S. of the linear effect, which was $\dfrac{(\sum \xi'_1 y)^2}{r \sum \xi'^2_1}$ for totals, would be $\dfrac{(\sum \xi'_1 y)^2}{\sum \xi'^2_1}$

REGRESSION

Because we had orthogonal contrasts, $\sum \xi_1' = 0$. In the present case we have a series of xs instead of ξ_1's and can turn them into similar coefficients by taking $(x - \bar{x})$ instead of ξ_1' and, using the same definitions as before, we expect the linear effect, which is the b of the expression $y = a + bx$, to be

$$\frac{\sum(x - \bar{x})y}{\sum(x - \bar{x})^2}$$

$\sum(x - \bar{x})$ must be zero as was shown on p. 23. The computations are simplified if we replace y by $y - \bar{y}$ also. This makes no difference to the value obtained since

$$\sum_{i=1}^{n}(x_i - \bar{x})(y_i - \bar{y}) = (x_1 - \bar{x})(y_1 - \bar{y}) + (x_2 - \bar{x})(y_2 - \bar{y}) + \ldots + (x_n - \bar{x})(y_n - \bar{y})$$

$$= (x_1 - \bar{x})y_1 + (x_2 - \bar{x})y_2 + \ldots + (x_n - \bar{x})y_n$$
$$- \bar{y}\{(x_1 - \bar{x}) + (x_2 - \bar{x}) + \ldots + (x_n - \bar{x})\}$$

and, since the terms inside the braces sum to zero,

$$= \sum(x - \bar{x})y$$

So our final model becomes $y = \bar{y} + b(x - \bar{x})$.

The expression $\sum(x - \bar{x})(y - \bar{y})$ is usually called the sum of products (S.P.). In the same way as a sum of squares divided by degrees of freedom estimates the variance of a population made up of single independent variates (or a *univariate population* to give it its statistical name) so S.P. divided by degrees of freedom estimates the co-variance (C_{xy}) in a *bivariate population*, i.e. a population in which there are independent pairs of observations, although the two members of any pair are not independent of each other.

Simple methods of obtaining sums of products exist and are similar to those for sums of squares:

$$\sum_{i=1}^{n}(x - \bar{x})(y - \bar{y}) = \sum xy - \bar{y}\sum x - \bar{x}\sum y + n\bar{x}\bar{y}$$

$$= \sum xy - \frac{\sum y \sum x}{n} - \frac{\sum x \sum y}{n} + \frac{n \sum x \sum y}{n^2}$$

$$= \sum xy - \frac{\sum x \sum y}{n}$$

Thus S.P. can be obtained by summing the actual products and subtracting as a "correction factor" the product of the two totals divided by the number of observations. It will be seen that putting x for y the formula reduces to the sum of squares for x.

If there are only two pairs of measurements S.P. becomes

$$\left(x_1 - \frac{x_1+x_2}{2}\right)\left(y_1 - \frac{y_1+y_2}{2}\right) + \left(x_2 - \frac{x_1+x_2}{2}\right)\left(y_2 - \frac{y_1+y_2}{2}\right)$$
$$= \tfrac{1}{4}(x_1-x_2)(y_1-y_2) + \tfrac{1}{4}(x_2-x_1)(y_2-y_1)$$
$$= \tfrac{1}{2}(x_1-x_2)(y_1-y_2)$$

which is similar to the expression of half the square of the difference for calculating the S.S. when there are only two values. All the other rules for orthogonal contrasts also hold, and the divisors can be worked out in exactly the same way as in the univariate case.

Returning to our example, we must first obtain the totals for x and y. $\sum x = 11 \cdot 86$ and $\sum y = 231 \cdot 3$.

S.S. for x = S.S.$_x$ = 0·4540

S.S. for y = S.S.$_y$ = 255·7794

S.P. for xy = S.P.$_{xy}$ = $0 \cdot 37 \times 8 \cdot 2 + 0 \cdot 59 \times 10 \cdot 6 \ldots + 0 \cdot 94 \times 23 \cdot 4$

$$- \frac{11 \cdot 86 \times 231 \cdot 3}{16} = 180 \cdot 215 - 171 \cdot 451 = 8 \cdot 764$$

Now the estimated linear effect, b in the model, is

$$b = \frac{\text{S.P.}_{xy}}{\text{S.S.}_x} = \frac{8 \cdot 764}{0 \cdot 4540} = 19 \cdot 3037$$

However, we want to have some degree of certainty that the true value is not zero, so we must test this estimate in the usual null-hypothesis form by saying we believe that the true b (usually given the symbol β by statisticians) is zero and see how frequently we could get a value as far away from zero as ours if this was true. To do this we need an estimate of error and, since there are no exact replications to give us the within treatments variation, we use the residual variation in the ys. This makes biological sense, in that the total S.S. for y measures the total variation in the experiment. If we subtract the S.S. for the linear effect, we are left with other effects which according to our model can only be the same as natural variation. It is important to realize this assumption when using this method. If b turns out to be not significant, it does not mean that there is no relationship, but that there is no *linear* relationship. For example, if $y = (x - \bar{x})^2$, the S.S. for the linear effect would be only natural variation, whereas the residual against which it is tested would contain all the S.S. for the quadratic effect and would be very large.

However, we should set out an analysis of variance of y:

Sources of variation	D.F.	S.S.	M.S.	F
Linear effect	1	169·1799	169·1799	27·35***
Residual	14	86·5995	6·1857	
Total	15	255·7794		

The S.S. for the linear effect is $\text{S.P.}^2_{xy}/\text{S.S.}_x$ as indicated earlier and in this case

$$\frac{8·764^2}{0·4540} = 169·1799$$

The residual or error S.S. is

$$\text{S.S.}_{\cdot y} - \frac{\text{S.P.}^2_{\cdot xy}}{\text{S.S.}_{\cdot x}} = 255·7794 - 169·1799 = 86·5995$$

The linear effect is a single contrast, so has 1 D.F. There were 16 observations of y, so the total D.F. is 15; and therefore the residual has $15 - 1 = 14$ D.F. Another way of looking at this biologically can be seen by referring to figure 7.5. The total S.S. represents the variation of the dots (actual values of y) around the mean y. The S.S. for the linear effect represents the variation of the value of y estimated by the line around the mean y; so, by difference, the residual S.S. represents the variation of the dots around the estimated values on the line. The D.F. for this residual also confirm a claim made on p. 28 that D.F. is the number of observations less the number of parameters estimated by the observations. For the total S.S. only one parameter, the mean y, was estimated but, for the residual, both mean y and true b have been estimated, so it has 16 observations minus 2 parameters, i.e. 14 D.F.

Mean squares can be calculated in the usual way and an F-test carried out, when we see that b is very significant. We could equally well test b by a t-test. Following the usual rules for linear functions, its S.E. will be

$$\sqrt{\frac{\text{M.S.}_E}{\sum k^2}} = \sqrt{\frac{\text{M.S.}_E}{\text{S.S.}_{\cdot x}}} = \sqrt{\frac{6·1857}{0·4540}} = \pm 3·6912 \quad \text{and} \quad t = \frac{19·3037}{3·6912} = 5·23$$

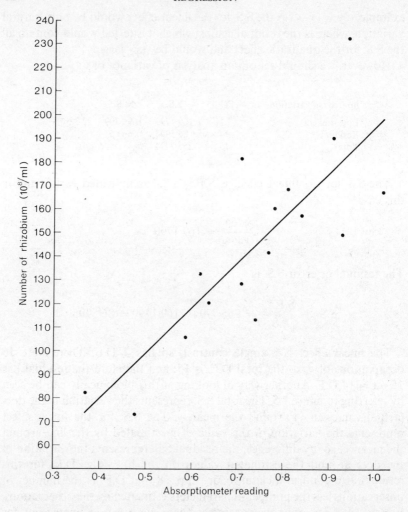

Figure 7.5—Regression for determining number of rhizobium from absorptiometer reading. ● observed values.

We can therefore go on and set confidence limits for our estimated *y*s. As far as the points on the curve are concerned, we follow again the usual rules for linear functions. Calling a point on the curve Y to avoid confusing it with an observed value y,

$$Y = \bar{y} + b(x - \bar{x})$$

so the S.E. for Y is

$$\sqrt{(\text{S.E.}_{\bar{y}}^2 + \text{S.E.}_{b(x-\bar{x})}^2)}$$

$$= \sqrt{\left(\frac{\text{M.S.}_E}{n} + \frac{(x-\bar{x})^2 \text{M.S.}_E}{\text{S.S.}_{\cdot x}}\right)}$$

$$= \sqrt{\left[\text{M.S.}_E \left(\frac{1}{n} + \frac{(x-\bar{x})^2}{\text{S.S.}_{\cdot x}}\right)\right]}$$

The confidence limits are $Y \pm t(\text{S.E.}_Y)$. It is seen that the S.E. is not the same for all Ys but varies according to the value of x. It is least when $x = \bar{x}$ and gets larger as the x gets further away from the mean, an important consideration for the biologist doing this kind of work. He should ensure that his area of interest is near the middle of any line he sets up, because it is there that he can be most confident of his prediction. At the ends of the line we are less confident and, if we try to extrapolate beyond the range we have investigated, our prediction becomes even less precise since then we have no evidence that the relationship still holds. It is very hazardous to extrapolate beyond the range that has actually been studied.

To use the function for determining number of cells from the absorptiometer readings we should first collect together the constant terms.

$$Y = \bar{y} + b(x-\bar{x}) \quad \text{becomes} \quad Y = (\bar{y} - b\bar{x}) + bx$$

$$\bar{y} = \frac{231 \cdot 3}{16} = 14 \cdot 4562; \quad \bar{x} = \frac{11 \cdot 86}{16} = 0 \cdot 7412; \quad b = 19 \cdot 3037$$

So $Y = 14 \cdot 4562 - 14 \cdot 3079 + 19 \cdot 3037x = 0 \cdot 1483 + 19 \cdot 3037x$.

Thus, if we observed on a particular suspension an absorptiometer reading of 0·60, our best estimate of the number of cells (Y) would be

$$(0 \cdot 1483 + 19 \cdot 3037 \times 0 \cdot 60) \times 10^8/\text{ml} = 11 \cdot 73 \times 10^8/\text{ml}$$

and the 95 per cent confidence limits are

$$11 \cdot 73 \times 10^8 \pm 2 \cdot 14 \sqrt{\left[6 \cdot 1857 \left(\frac{1}{16} + \frac{(0 \cdot 60 - 0 \cdot 7412)^2}{0 \cdot 454}\right)\right]} \times 10^8/\text{ml}$$

$$= 11 \cdot 73 \times 10^8 \pm 1 \cdot 74 \times 10^8/\text{ml}$$

or we can be 95 per cent confident that the true number of cells in that suspension lies between $9 \cdot 99 \times 10^8$ and $13 \cdot 47 \times 10^8/\text{ml}$.

There may be occasions when we would want to make a statement about the value of y we would expect to observe in a single experiment a specified x, and not about the average y for this x as above. example, at a later date we might measure both x and y on a ticular suspension and wish to know whether this pair of values gives ny cause for doubting the reliability of our line. In terms of confidence limits, we should wish to know within what limits y for a single sample with given x might lie. We then require an S.E. for y observed $-Y$ estimated

$$= \sqrt{(\text{S.E.}_y^2 + \text{S.E.}_Y^2)}$$
$$= \sqrt{\left[\text{M.S.}_E + \text{M.S.}_E\left(\frac{1}{n} + \frac{(x-\bar{x})^2}{\text{S.S.}_x}\right)\right]}$$
$$= \sqrt{\left[\text{M.S.}_E\left(1 + \frac{1}{n} + \frac{(x-\bar{x})^2}{\text{S.S.}_x}\right)\right]}$$

Such confidence limits are of course much wider than those above.

It will be noticed that, in this analysis, the variation used for the test is variation in y, so the distribution of y and the variance of y must reasonably fit the assumptions implied in an F-test; but there are no restrictions on the distribution of the xs. Thus we can use this method when we deliberately choose each x, as when setting up a range of dilutions in a biochemical problem, or when selecting individuals more or less at random to cover a range. Also it can be used when two measures arise naturally, e.g. number of grains per ear and weight per grain. Then it is quite reasonable to ask: how is weight per grain affected by number of grains per ear, on the assumption that there is genetic control of the number of grains set, and these must then share a limited amount of assimilates. It is equally reasonable to ask: what is the effect of weight per grain on number of grains per ear, if we consider that grains are filled in sequence and, when all assimilates are used up, no more grains will be filled. However, in all cases, we use regression when we want to find the effect of one variable x, called the *independent* variable, on another variable y, called the *dependent* variable. Which is x and which is y should be controlled by the null hypothesis made, and regression will give precise answers to such hypotheses.

Correlation

Regression differs from another statistical procedure known as *correlation*, which uses similar arithmetic. This simply answers the question: are these two variables x and y associated? i.e. as one changes the other changes too. This knowledge may be useful at the very beginning of a biological investigation, but it does not lead very far. Suppose the answer is yes. There are at least three possible reasons: (1) change in x causes change in y, (2) change in y causes change in x, (3) change in some other factor z causes change in x and in y at the same time. Sometimes we can rule out one of these possibilities for biological reasons, but it is seldom possible to rule out two. In the grain-filling example (p. 126), all are possible; but if we found association between leaf area and weight per grain, we would not imagine that weight per grain could affect leaf area, since the latter is determined before the grains are filled. But there are still two distinct possibilities: (1) the larger leaves produce more assimilates and therefore heavier grain, but (2) some other factor, like potassium supply, increases leaf area and increases size of grain. If we accepted the first, and set about increasing leaf area by applying nitrogen, we should be wasting our time if the second was true.

However, to show the arithmetic involved we will use the results of the absorptiometer tests on p. 120. The *correlation coefficient* is defined as

$$r = \frac{\text{S.P.}_{xy}}{\sqrt{(\text{S.S.}_x \text{S.S.}_y)}}$$

It is interesting to see how it is connected with regression. The b for the regression of y on x, or the way x affects y (call it b_{yx}) was $\text{S.P.}_{xy}/\text{S.S.}_x$. If we worked out b for the regression of x on y, i.e. the way y affects x, b_{xy} would obviously be

$$\frac{\text{S.P.}_{xy}}{\text{S.S.}_y} \quad \text{so} \quad b_{yx} b_{xy} = \frac{\text{S.P.}_{xy}^2}{\text{S.S.}_x \text{S.S.}_y}$$

and therefore

$$r = \sqrt{(b_{yx} b_{xy})}$$

in words the geometric mean of the two regression coefficients. In the present example

$$r = \frac{8 \cdot 764}{\sqrt{(0 \cdot 4540 \times 255 \cdot 7794)}} = 0 \cdot 813$$

Theoretically r can vary from zero, when there is no association, to ± 1, when there is complete agreement. Provided the usual assumptions concerning distributions and variances hold (in this case both x and y must be normally distributed) the null hypothesis that the true value of r (usually given the symbol ρ) is zero can be tested by a t-test where

$$t = \frac{r\sqrt{(n-2)}}{\sqrt{(1-r^2)}} = 5.22$$

is referred to the table (Appendix Table II) with t of $(n-2)$ D.F. Tables for testing $\rho \neq 0$ directly are available in Fisher and Yates (1963).

It must be remembered when interpreting r that, like b, it will only detect linear relationships, and the two variables may well be associated in some other way.

There is one other connection between r and b which is sometimes useful to biologists. In the analysis of variance on p. 123 the regression S.S. was calculated as $S.P._{xy}^2/S.S._x$ and the total S.S. was $S.S._y$, so the proportion of the sums of squares accounted for by the regression $= S.P._{xy}^2/(S.S._x \, S.S._y) = r^2$. This proportion, usually expressed on a percentage scale by multiplying r^2 by 100, is often called the *coefficient of determination* and is a general measure of the usefulness of a regression.

Kendall's rank correlation

If the variables measured in an experiment do not conform with the assumptions for a test of correlation by the method just described, there are non-parametric tests available. One such is to calculate Kendall's coefficient of rank correlation. Suppose we had observed the chiasma frequency per cell and fertility in plants of an autotetraploid rye. We might wish to know if there was association between the number of univalents (a single cross of two chromosomes) (x) and fertility (y), but it is unlikely that the number of univalents would be normally distributed. To make the test we first set out the data in rank order of one of the variables. (Here we take x, frequency of univalents, but it does not matter which.)

x	0	0.05	0.10	0.15	0.20	0.225	0.25	0.40	0.50	0.60
y	19.5	40.7	39.4	25.7	26.4	24.4	23.0	19.6	22.6	9.28
P	8	0	0	1	0	0	0	1	0	
Q	1	8	7	5	5	4	3	1	1	

Then working on the variable we did not rank (y, fertility, in this case) and starting with the extreme left-hand value, we count the number of values to the right of it which exceed it (P) and the number of values which are less than it (Q). This gives $P = 8$, $Q = 1$ in this case. Now we repeat this process with each y-value in turn. We sum the Ps and the Qs (10 and 35 in this case); as a check

$$\sum P + \sum Q = \tfrac{1}{2}n(n-1)$$

and $\sum P - \sum Q$ gives the test statistic (usually given the symbol S). In this case $S = 10 - 35 = -25$. Kendall's rank correlation coefficient is defined as

$$\tau = \frac{2S}{n(n-1)}$$

which in this case is

$$\frac{-50}{10 \times 9} = -0.556$$

As with r, theoretically τ can vary from -1 to $+1$, $+1$ if the order of x and y is exactly the same, -1 if the orders are exactly opposite. It must be remembered that it gives only a general indication of the closeness of the relationship; it does not tell us anything about the form of the relationship, in particular $\tau = 0$ can occur when x and y are completely unrelated, but also when they have a very close curved relationship such as $y = x^2$. There are tables for testing S directly (Siegel, 1956, p. 285) and in this case a value -25 or less can occur with a probability of 0·014. Since we had no reason to specify that τ should be negative, we should also consider the probability of $+25$ or a greater number occurring. We would conclude that if the true $\tau = 0$ we have witnessed an event with a probability of only 0·028, so would probably be satisfied that there is association between the number of univalents and fertility in this rye.

We have considered linear regression, which is the form most commonly used in biology, but the same principles apply to any other form of relationship. Similar procedures of dividing the total sum of squares into two parts, one due to the curve which is fitted, and the other to the deviations from that curve, are used. As with testing contrasts or differences between particular treatments, the type of curve must be defined in the null hypothesis. To plot the points, choose a likely looking function, and then test it on the same data and claim that this function is true for anything other than that sample, is pure self-deception.

CHAPTER EIGHT

ERROR VARIATION AND ITS CONTROL

The one-way classification design which was used in all the illustrations of choosing treatments in Chapters 5, 6 and 7 can be looked upon as based on the model specified on p. 48:

$$x_{ij} = m + a_i + e_{ij}$$

where x_{ij} is the jth value of the ith treatment, m is a constant applying to the particular material being studied, a_i is the effect of the ith treatment, and e_{ij} is the natural variation attached to this particular value of x and not otherwise accounted for by the model. We estimate the value of m as the general mean of the whole experiment, and a_i as the difference between the mean of all the values for the ith treatment and the general mean.

Let us suppose that we have an experiment with p treatments each with r replications, so that i takes values from 1 to p and j values from 1 to r. We would calculate the mean square for treatments as

$$\frac{\text{S.S.}}{p-1}$$

where S.S. was calculated as

$$\frac{T_1^2 + T_2^2 + \ldots + T_p^2}{r} - \frac{(\text{grand total})^2}{rp}$$

where the Ts are treatment totals.

Substituting the values expected from the model we have

$$\text{M.S.} = \frac{1}{p-1} \left[\frac{1}{r} \{(rm + ra_1 + e_{11} + e_{12} + \ldots + e_{1r})^2 \right.$$
$$+ (rm + ra_2 + e_{21} + e_{22} + \ldots + e_{2r})^2 + \ldots$$
$$+ (rm + ra_p + e_{p1} + e_{p2} + \ldots + e_{pr})^2\} - \frac{1}{rp}$$
$$\left. \times (rpm + ra_1 + ra_2 + \ldots + ra_p + e_{11} + e_{12} + \ldots + e_{1r} + e_{21} + \ldots + e_{pr})^2 \right]$$

It is immediately apparent that terms containing m disappear on simplifying this expression, e.g. we have

$$\frac{pr^2m^2}{r} - \frac{r^2p^2m^2}{rp} = 0$$

To simplify the rest of the expression, and find what its value would be on average over a large number of experiments, i.e. its expectation in the mathematical sense, we have to make a number of assumptions. First we can assume (or define) as on p. 48 that the parts of the model are independent, i.e. each a is independent of any other a and of any e, and the es are independent of each other; then the expectation of all product terms like $a_i a_j$, $a_i e_{ij}$ and $e_{ij} e_{kl}$ is zero. We are left then with the a^2 and e^2 terms and these sum to

$$\frac{1}{p-1}\left(\frac{r^2 \sum a^2}{r} - \frac{r^2 \sum a^2}{rp}\right) = \frac{1}{p-1} \cdot \frac{r(p-1)}{p} \sum a^2 = \frac{r \sum a^2}{p}$$

and

$$\frac{1}{p-1}\left(\frac{\sum e^2}{r} - \frac{\sum e^2}{rp}\right) = \frac{1}{p-1} \frac{(p-1)}{rp} \sum e^2 = \frac{\sum e^2}{rp}$$

Now $(\sum e^2)/rp$ is our usual estimate of error variance, σ^2, but we still have two possible definitions of $(\sum a^2)/p$.

If we have treatments which are defined entities that we want to make remarks about, such as three amino acids, arginine, methionine and lysine, we call treatments a *fixed-effect factor* and consider $(\sum a^2)/p$ to measure the variation between the levels of that factor, hence a measure of the effect of the factor, giving it the symbol A^2. If, however, the levels of our treatment factor are just a random selection of the things with that name, like a selection of habitats from which a particular species has been obtained, then the treatments are considered to be a *random-effect factor* and $(\sum a^2)/p$ estimates the variance of the population of such treatments and is given the symbol σ_A^2.

A good biological test for deciding whether treatments represent a fixed- or random-effect factor is to think how we would view the situation if one of the levels was changed. If this made it a different experiment, e.g. if we substituted tyrosine for arginine in the amino acid experiment, thereby testing something quite different, then we have a fixed-effect factor. If, on the other hand, we considered it the same experiment, as we would when collecting plants from a random selection of habitats, when two different selections would be considered to be finding out the same thing, then we have a random-effect factor. It is important to decide, not only from the testing point of view, but also to be sure of the conclusions that can be drawn.

Thus the two possible models for the analysis of variance of the single factor design are:

| | | Expected mean square | |
Source of variation	D.F.	Fixed-effect factor	Random-effect factor
Treatments	$p-1$	$\sigma^2 + rA^2$	$\sigma^2 + r\sigma_A^2$
Error	$p(r-1)$	σ^2	σ^2

and it is seen how an F-test can operate. If the null hypothesis is true, A^2 or σ_A^2 is zero, leading to an expected F-value of 1; so if F is much greater than 1, we find it difficult to believe that A^2 or σ_A^2 is zero. Tables based on the frequency of occurrence of various F-values when the null hypothesis is true give a measure of confidence for adopting this disbelief.

Two useful points of principle come out of these calculations.

(1) When a sum of squares is derived by squaring totals, each of r individuals, then the mean deviations of the effects appear in the M.S. expression as $(ra)^2/rp$ because the mean deviation is multiplied by r to obtain the total, and the squares of the totals are divided by r to obtain the S.S. for reasons given earlier. Thus, if $(\sum a^2)/p$ estimates the variance σ_A^2, the contribution to the expected mean square will be $r\sigma_A^2$, and this forms a universal rule:

When all levels of a factor have the same replication, its contribution to the expected mean square can always be written down as the variation due to that factor multiplied by the number of individuals used to form the total for each level.

This enables us to get estimates of variance for particular factors, e.g. from the random-effect column of the above analysis of variance we see that $r\sigma_A^2 + \sigma^2$ is estimated by the M.S. for treatments and σ^2 by the M.S. for error, and since

$$\sigma_A^2 = \frac{r\sigma_A^2 + \sigma^2 - \sigma^2}{r}$$

it is estimated by

$$\frac{\text{M.S. for treatments} - \text{M.S. for error}}{r}$$

(2) We see that to make a proper test the error variation must contain

ERROR VARIATION AND ITS CONTROL 133

all the variation contained in the factor being tested, except for the one item appropriate to the factor alone.

Hierarchical design

We can now consider an example which illustrates a type of design used frequently for genetical purposes of estimating variances and known as a *hierarchical* or *nested design*. In a breeding experiment designed to determine the variability of a certain strain of pigs, both sires and dams were chosen at random, and the dams randomized to sires at mating. The progeny (of which only three were taken from each litter in this case— probably all of one sex) provided the data. The results are shown in Table 8.1.

Table 8.1—Results of a breeding experiment with pigs.
Average daily gain (lb.)

Sire	Dam	Pig gains (x)			(D)	Totals (S)	
A	1	2·77	2·38	2·56	7·71		
	2	2·79	3·14	2·95	8·88	16·59	
B	3	2·08	2·02	2·06	6·16		
	4	3·01	2·61	2·81	8·43	14·59	
C	5	2·36	2·71	2·52	7·59		
	6	2·72	2·74	2·73	8·19	15·78	
D	7	2·87	2·46	2·60	7·93		
	8	2·11	2·04	2·08	6·23	14·16	
E	9	2·84	2·66	2·77	8·27		
	10	2·40	2·38	2·39	7·17	15·44	76·56

Analysis of variance

Source of variation	D.F.	S.S.	M.S.	Parameters estimated
Sires	4	0·6195	0·1549	$\sigma^2 + 3\sigma_D^2 + 6\sigma_S^2$
Dams within sires	5	1·8303	0·3661	$\sigma^2 + 3\sigma_D^2$
Pigs within dams	20	0·3871	0·0194	σ^2

To analyse, we first obtain the totals for each dam, then add the total for dams to get totals for sires. The D.F. are $(5-1) = 4$ for sires, $5 \times 1 = 5$ for dams within sires, and $10 \times 2 = 20$ for pigs within dams, summing to 29, as we would expect since we have 30 pigs. We then calculate the

$$\text{C.F.} = \frac{76 \cdot 56^2}{30} = 195 \cdot 3811$$

Then sires S.S. will be

$$\frac{16 \cdot 59^2 + 14 \cdot 59^2 + \ldots + 15 \cdot 44^2}{6} - \text{C.F.} = 196 \cdot 0006 - 195 \cdot 3811 = 0 \cdot 6195$$

or, in general, if there are d dams per sire and p pigs per dam

$$\frac{\sum S^2}{dp} - \text{C.F.}$$

Now S.S. for dams within sires is the sum of the sums of squares of dams within each sire, e.g. within A we have

$$\frac{7\cdot71^2 + 8\cdot88^2}{3} - \frac{16\cdot59^2}{6}$$

So when we add them up we get

$$\frac{7\cdot71^2 + 8\cdot88^2 + 6\cdot16^2 + 8\cdot43^2 + \ldots + 7\cdot17^2}{3} - \frac{16\cdot59^2 + 14\cdot59^2 + \ldots + 15\cdot44^2}{6}$$

$$= 197\cdot8309 - 196\cdot0006 = 1\cdot8303$$

or in general

$$\frac{\sum D^2}{p} - \frac{\sum S^2}{dp}$$

Likewise pigs within dams will be

$$2\cdot77^2 + 2\cdot38^2 + \ldots + 2\cdot39^2 - 197\cdot8309 = 0\cdot3871 \quad \text{or} \quad \sum x^2 - \frac{\sum D^2}{p}$$

We can then fill in the M.S. column by dividing S.S. by D.F. to get 0·1549, 0·3661 and 0·0194.

The expected M.S. or parameters estimated are obtained by a simple extension of the one-way classification and starting from the bottom will be:

σ^2, the basic pig variation;
$\sigma^2 + 3\sigma_D^2$ (3 because there are three pigs contributing to each value that was squared); and
$\sigma^2 + 3\sigma_D^2 + 6\sigma_S^2$ (6 because there are six pigs in each sire total).

We might now test if there really is any variation due to dams by using an F-test of

$$\frac{0\cdot3661}{0\cdot0194} = 18\cdot87$$

with 5, 20 D.F. so the evidence against $\sigma_D^2 = 0$ is very strong indeed. To test if the variation due to sires is anything other than zero, we must consider

F derived from

$$\frac{\text{M.S. for sires}}{\text{M.S. for dams within sires}}$$

since it is the M.S. for dams within sires which contains all that is in the M.S. for sires except the sire effect. The M.S. of pigs within dams would not be appropriate, because we could get significance simply because σ_D^2 was large whilst σ_S^2 was zero. So the appropriate F-value for sires is

$$\frac{0{\cdot}1549}{0{\cdot}3661}$$

which is less than one with 4, 5 D.F., and we would see no reason for believing that σ_S^2 is anything other than zero.

To obtain an estimate of σ_D^2 we have

$$\frac{\text{M.S. for dams within sires} - \text{M.S. for pigs within dams}}{3}$$

$$= \frac{0{\cdot}3661 - 0{\cdot}0194}{3} = 0{\cdot}1156$$

Thus both tests and estimates of variances can be obtained from analysis of variance, but sometimes we need to set confidence limits for variances also. Variances from normal populations do not themselves follow a normal distribution, so we cannot use t to set confidence limits. However, if s^2 is our estimate of a true variance σ^2, with $(n-1)$ D.F. then

$$\frac{(n-1)s^2}{\sigma^2}$$

follows approximately the χ^2-distribution, which was introduced on p. 62. If

$$\frac{(n-1)s^2}{\sigma^2} = \chi^2 \quad \text{then} \quad \sigma^2 = \frac{(n-1)s^2}{\chi^2}$$

χ^2-tables are two-dimensional with D.F. and the probability of obtaining a χ^2 as large as the value given, so following the argument used for setting confidence limits using t, we can write for confidence limits of probability α

$$\frac{(n-1)s^2}{\chi^2_{\frac{1}{2}(1-\alpha)}} \leqslant \sigma^2 \leqslant \frac{(n-1)s^2}{\chi^2_{\frac{1}{2}(1+\alpha)}}$$

where $\chi^2_{\frac{1}{2}(1-\alpha)}$ and $\chi^2_{\frac{1}{2}(1+\alpha)}$ are the tabular values of χ^2.

Thus, for variation of dams, the 95 per cent confidence limits are

$$\frac{5 \times 0 \cdot 1156}{12 \cdot 83} \leqslant \sigma_D^2 \leqslant \frac{5 \times 0 \cdot 1156}{0 \cdot 83}$$

where for 5 D.F. Appendix Table III shows that the 0·025 and 0·975-values for χ^2 are 12·83 and 0·83, respectively. So we are 95 per cent confident that σ_D^2 lies between 0·0451 and 0·6964. It is noticeable that these limits are not equidistant from the point estimate of 0·1156. This is because, unlike the t-distribution, the χ^2-distribution is very skewed; hence the cut-off areas have to be calculated separately for each end.

Two-factor design

Experiments with more than one factor can also be considered in the form of a model similar to the one used for the single-factor situation. Suppose there were two factors A and B, and we included all combinations of their levels, say p and q respectively, and had r replications of each combination. The model would be $x_{ijk} = m + a_i + b_j + (ab)_{ij} + e_{ijk}$ where m is again a constant referring to the experimental material and estimated by the general mean; a_i the effect of the ith level of the factor A and estimated as the difference between the mean of all members receiving a_i and the general mean; b_j would be defined similarly for the factor B; $(ab)_{ij}$ would be the interaction effect appropriate to the members receiving the treatment $a_i b_j$, and e_{ijk} would be the natural variation element associated with the particular x_{ijk}. When deciding upon the components represented in the various mean squares, we have to make one or other of the two assumptions concerning the nature of the effects for each factor separately. Thus both A and B may be considered fixed-effect factors, both may be random-effect factors, or one may be fixed and the other random, giving what is called a *mixed model*. In addition, we have the interaction effect and, if both A and B are fixed-effect factors, then the interaction AB will itself be a fixed effect. If one or both of them is a random-effect factor, we assume that the interaction is a random effect. The difference between a fixed-effect interaction and a random-effect one in manipulating the model is that for the former we assume $\sum (ab)_{ij} = 0$ for each level of A and for each level of B, whereas for the latter we assume that the $(ab)_{ij}$ are distributed normally with a mean of a zero and a variance of σ_{AB}^2. We again make the assumption of independence, so that cross-product terms sum to zero, and when expressions for mean squares have been derived, find again that all terms containing m cancel out, but also that in the expression for A, all the

b-terms cancel out as do the a-terms in the expression for B. The final expected mean squares turn out to be:

Expected mean squares

Source of variation	D.F.	A and B both fixed effects	A and B both random effects	A a random effect B a fixed effect
A	$p-1$	$\sigma^2 + qrA^2$	$\sigma^2 + r\sigma_{AB}^2 + qr\sigma_A^2$	$\sigma^2 + r\sigma_{AB}^2 + qr\sigma_A^2$
B	$q-1$	$\sigma^2 + prB^2$	$\sigma^2 + r\sigma_{AB}^2 + pr\sigma_B^2$	$\sigma^2 + r\sigma_{AB}^2 + prB^2$
AB	$(p-1)(q-1)$	$\sigma^2 + r(AB)^2$	$\sigma^2 + r\sigma_{AB}^2$	$\sigma^2 + r\sigma_{AB}^2$
Error	$pq(r-1)$	σ^2	σ^2	σ^2

Thus it is seen that, when both factors are fixed-effect factors, there must be some additional form of replication to provide an error mean square for testing any of the effects but, if one of the two factors is a random-effect factor, then the interaction forms the appropriate error for testing the factors and, only if the interaction itself requires testing, is it necessary to have further replication.

This idea leads to some very useful experimental designs for biologists, because very often we wish to test effects over a wide range of conditions but have no particular wish to define these conditions, further than to say that they come from a population of conditions that can be found naturally.

For example, a research student wanted to compare the time that two different species of aphid would stay on a potato plant before migrating, because he had the null hypothesis that the two species did not differ in time spent on a plant. Not much was known about the variation within the two populations of aphids, but it was known that the time that an aphid will spend on a leaf of a potato plant depends on the age of the leaf. Also, from the point of view of recognition it would not be possible to have a large number of aphids on any one leaf, as there would be rather imprecise records of who left when. The answer was to have two aphids, one of each species, on each of several leaves, so 15 suitable leaves were chosen and numbered 1 to 15. Potato leaves have a mid-rib which is a very clear mark down the middle, and it was decided to put one aphid on each side of this mid-rib, one species a_0 on one side and the other a_1 on the other side, the positions being chosen by random numbers.

To do this we simply randomize which species shall be on the left-hand side of the mid-rib of each leaf, and the other species must be on the right-hand side of that leaf. The aphids were then put on the leaves in order, putting the appropriate species on the left-hand side of leaf 1, then the other on its right-hand side, then the same for leaf 2 and so on. This could be important if it took a long time to put them on, since, if all the a_1s were put on first, differences between species might be confused with the different times of day at which the experiment started. Then the experimenter sat and watched, and recorded the time when each aphid left. The results are given in Table 8.2.

Table 8.2—Results and analysis of an experiment on the migratory habits of aphids

Layout

Leaf number

	1	2	3	4	5	6	7	8	9	10	11	12	13	14	15
Left-hand side	a_1	a_1	a_1	a_1	a_1	a_0	a_0	a_1	a_1	a_0	a_1	a_1	a_0	a_1	a_0
Right-hand side	a_0	a_0	a_0	a_0	a_0	a_1	a_1	a_0	a_0	a_1	a_0	a_0	a_1	a_0	a_1

Results (time spent in minutes)

Leaf number

	1	2	3	4	5	6	7	8	9	10	11	12	13	14	15	Total
a_0	280	289	287	246	268	308	338	285	269	246	317	276	329	265	356	4359
a_1	260	209	196	213	193	210	240	166	309	189	316	172	242	153	327	3395
Total(L)	540	498	483	459	461	518	578	451	578	435	633	448	571	418	683	7754
Diff. (d)	+20	+80	+91	+33	+75	+98	+98	+119	−40	+57	+1	+104	+87	+112	+29	+964

Analysis of variance

Source of variation	D.F.	S.S.	M.S.
Leaves	14	42 779	
Species	1	30 977	30 977
Error (leaves × species)	14	14 915	1065
Total	29	88 671	

The degrees of freedom are 14 for leaves, since there are 15 leaves in the experiment, 1 for the difference between the two species and $14 \times 1 = 14$ for the interaction between leaves and species. If we assume that leaves represent a random-effect factor, this is the appropriate error term for testing the difference between species. Our biological test for type of effect works here; the research student would not have deliberately chosen the leaves and would be perfectly happy with the result if his assistant had lost all the leaves he had gathered and he had to get some more.

The S.S. for leaves can be calculated from the leaf totals as

$$\frac{\sum L^2}{2} - \frac{(\sum x)^2}{2 \times 15} = \frac{4\,093\,860}{2} - 2\,004\,151 = 42\,779$$

Since species has only two levels, its sum of squares can be calculated from the difference between the two totals, i.e.

$$\frac{(4359-3395)^2}{2\times 15} = \frac{964^2}{30} = 30\,977$$

The interaction or error sum of squares can also be calculated directly when one factor has but two levels as shown on p. 115. In this case we calculate the difference between a_1 and a_0 for each leaf (d in Table 8.2) and the S.S. $= \frac{1}{2}\sum d^2 - (\sum d)^2/(2\times 15)$. Noticing that $(\sum d)^2/(2\times 15)$ is the S.S. for species that we have already worked out, S.S. for error

$$= \frac{91\,784}{2} - 30\,977 = 14\,915$$

As a check we would work out the total S.S. as

$$\sum x^2 - \frac{(\sum x)^2}{30} = 2\,092\,822 - 2\,004\,151 = 88\,671$$

which is exactly the same as the total of the three components calculated separately.

An F-test or a t-test would be made in the usual way, using the interaction mean square as the error term in the test.

This is the simplest form of what is probably the most useful design in experimentation. It is a two-factor design or more usually called *randomized-block design*, and is appropriate whenever it is possible to group the material into more homogeneous groups, and apply treatments equally within each group. A look at the analysis-of-variance table will show why it is superior to the one-way classification in this instance. If we had used the one-way classification, we should have considered that there were 30 positions on the leaves, and allocated the aphids entirely at random to the 30 positions. Thus the error sum of squares would have included the leaf variation as well as the leaf × treatment interaction which, of course, means that a much larger difference would be necessary for significance. By using the randomized-block design we have removed the effects of age of leaf from the treatment effects, because each treatment is affected equally by each leaf, and from the error because the variation due to age of leaf has been calculated separately.

The randomized-block design is very commonly used in biological experimentation, because there are many situations where it is possible to make up homogeneous groups while still exploring a fairly diverse population. In field experiments, small areas of land close together are usually more similar than the same-size areas scattered over a whole field. Thus an experimenter can often see how fertility changes and can divide the experimental area, first into blocks of similar fertility, and then randomize the treatments so that each block contains each treatment once and once only. In experiments with plants, they may first be grouped according to size or physiological age, or when growing plants during experiments, environmental effects may be taken care of by forming blocks of plants, e.g. blocks may be rows of plants in a glasshouse, thereby controlling the effects of nearness of the glass. Even in some of the best controlled-environment cabinets, there are environmental effects due to the points of entry of air and due to the difficulty of getting uniform light intensity, but such difficulties can be overcome by forming small blocks.

In experiments with small animals, positional effects of the cages can often be a source of great variation. If the cages are in tiers, treating each tier as a block will often allow much smaller differences to be detected.

Genetic effects can often be controlled in this way, by taking each litter as a block or, if the pedigree of the experimental animals is unknown, some control of variation can often be achieved by forming blocks according to initial live weight or age. In all experiments, the effects of time can be controlled by treating each occasion as a block. Thus a very expensive experiment, designed to test the effect of various feeding regimes on the production of beef by various breeds of cattle, was made possible by starting one complete replicate each year for 5 years, and treating years as blocks in the analysis of variance. On a smaller scale, we often find it impossible to carry out the experimental work on more than one replicate in one day, and again deliberately designing the experiment with "day" as a factor and analysing as a randomized-block design will often be advantageous.

When forming blocks in this way, there is no reason why several effects should not be combined or confounded, because there will be no particular interest in the differences between blocks. For instance, in a sheep experiment, one block might contain Border Leicester × Welsh sheep 10 months old which have been suckled by their dams, whilst another contains Finnish Landrace sheep 12 months old and artificially reared. When it came to shearing time, and there was insufficient labour to shear all the experimental animals in one day, we might well shear the Border Leicester × Welsh on a Friday and the Finns on the following Wednesday. Thus, a multiplicity of variation would be included in the block mean square, but excluded from the treatment and error mean squares leading to a more precise experiment.

This idea leads many experimenters to form blocks at the beginning of the experiment for no good reason, by some such method as deciding that the first member chosen at random for each treatment is in block I. They claim then that, if the need arises later, they are better equipped "to dump any extraneous variation" such as time effects into the blocks. However, if no such need arises, they have a less precise experiment, because for a given amount of replication the error degrees of freedom will be fewer for a randomized-block design than for a single-factor design with simple randomization $[(p-1)(r-1)$ compared with $p(r-1)]$ and, if blocks have no effect, the M.S. for error will be substantially the same in each case. Furthermore, if an accident occurs, and the experiment ends up with unequal replication, the calculations become more difficult with a randomized-block design than with a single-factor arrangement. Thus, the best advice seems to be to form blocks whenever it can be seen that there are real block effects, but otherwise start with the single-factor arrangement. If it is desirable to group the material later for time of an operation, say, form the blocks at that stage by allocating one individual of each treatment to each block at random.

Wilcoxon signed-rank test

If the assumptions for a t- or F-test cannot be met, the randomized-blocks concept can still be used with certain non-parametric tests. If the treatment factor has only two levels, we can use the Wilcoxon signed-rank test if the conditions are similar to those required for the Mann-Whitney (p. 49) and Kruskal-Wallis (p. 61) tests, i.e. if the two populations can be assumed to be of the same shape. If we cannot assume that, we can use the sign test which is applicable under almost any conditions. As in a design destined for an F- or t-test, we should randomize the treatments to the material, so that each treatment appears once and once only in each block.

For example, in an experiment to assess the virulence of two races of *Ps. phaseolicola* causing halo blight of beans, 10 plants of *Phaseolus coccineus* were taken at random and the first trifoliate leaflets inoculated with the bacteria, one race to each outer leaflet at random. After 10 days the extent of the resulting chlorosis was estimated. The data obtained are given in Table 8.3.

To use the Wilcoxon signed-rank test we first obtain the differences between races for each plant; taking race A as positive this gives column 4 of the table. Then these differences are ranked irrespective of sign to give column 5 (in this process, zero differences are ignored). We now sum

Table 8.3—Results of an experiment on halo blight Chlorosis (cm^2)

Plant	Race A	Race B	Difference (A − B)	Rank	Sign
1	15·6	14·3	+1·3	5	+
2	17·8	16·7	+1·1	4	+
3	13·2	14·1	−0·9	3	−
4	4·1	4·3	−0·2	1	−
5	5·5	2·2	+3·3	7	+
6	2·0	2·3	−0·3	2	−
7	8·7	4·0	+4·7	9	+
8	6·7	2·4	+4·3	8	+
9	7·5	1·7	+5·8	10	+
10	18·0	15·6	+2·4	6	+

separately the ranks which belong to positive differences and those which belong to negative differences (the signs are given in column 6). In this case, the sum of the ranks for positive differences, $R_+ = 5+4+7+9+8+10+6 = 49$ and for negative differences, $R_- = 3+1+2 = 6$. As with most ranking procedures, there is a simple check; since the first n natural numbers have been used for ranking, $R_+ + R_- = \frac{1}{2}n(n+1) = 5 \times 11 = 55$, which is right.

Tables have been prepared showing the probability of the more extreme values of R_- and R_+ for n non-zero differences up to 20 when the two populations are the same (Dixon and Massey, 1957, p. 443). In the present case with $n = 10$, the one-tail probability is 0·014. Since we have no prior hypothesis that Race A should be more virulent than Race B, but only the null hypothesis that they do not differ in virulence, $R_+ = 6$ and $R_- = 49$ is equally unlikely; so we double this probability (0·028) and can claim that, if the two races have the same virulence, we have witnessed an event which can occur in only 2·8 per cent of such trials; so would prefer to believe that they have not. The data in Table 8.3 emphasize the value of a randomized-block arrangement in this case. It is clear that the plants are very variable in their response to *Ps. phaseolicola*, no matter which race is used. This variability would cause a very high "error" if races had been allocated to whole plants at random, thereby making it impossible to detect any difference between the races.

Sign test

The sign test is even easier, though it is not so powerful since it uses less of the information. We simply count the number of times each treatment is greater than the other. In Table 8.3, Race A is greater than Race B on plants 1, 2, 5, 7, 8, 9 and 10, i.e. on 7 occasions, whilst Race B is greater than Race A on plants 3, 4 and 6, i.e. on 3 occasions (the sum of these two must obviously equal the number of non-zero differences). If the null hypothesis is true, these values should follow a very simple distribution called the *binomial distribution*. Tables have been constructed (e.g. Siegel, 1956, p. 250) and in the present case, if there is no difference, the probability of obtaining anything as extreme as 7 to 3 in favour of a particular race is 0·172, so the probability of such a difference is 2×0.172 or 0·344. Using this test, few people would be convinced that there is a true difference between the races. A look at the data shows why the two tests give such different interpretations; whenever Race B is the greater, the difference is very small, whereas Race A is appreciably greater than Race B on several plants. The signed-rank test takes some account of the size of the differences, whilst the sign test gives as much weight to the 0·2 increase due to B on plant 4 as it does to the 5·8 increase due to A on plant 9.

Generally to get any reasonable precision from a sign test, the number of comparisons must be large. It is worth noting that for n of 5 or less it is not possible to be 95 per cent confident of a true difference, even if one treatment is greater than the other on all occasions. The great merit of the test is that it does not require that numerical data are available, but can be used wherever a meaningful distinction can be made, e.g. a leaf is darker than another, an orchid is more beautiful than another, or one beer tastes better than another.

Friedmann procedure

If there are more than two treatments in a randomized-block design, Friedmann's non-parametric test is available. For example, it was suggested that consumers show no consistent preference for beef from particular breeds. Five members of the public were chosen at random, presented in a random order, chosen independently for each, with beef from five breeds (C, D, H, J, S) and asked to score for taste on any scale they thought appropriate, provided low values indicated least preferred and high values most preferred. The results are shown in Table 8.4.

The tasters would be equivalent to blocks, and the breeds of cattle the treatments. We should first rank the scores for each block separately, i.e. for taster no. 1 the lowest score is given to J which gets rank 1, C is next

Table 8.4—Scores for preference

Taster										
1	S	5	H	7	J	0	C	4	D	9
2	C	$2\frac{1}{4}$	S	2	H	$2\frac{1}{2}$	J	1	D	3
3	H	$7\frac{1}{2}$	D	9	J	6	C	$6\frac{1}{4}$	S	$6\frac{1}{2}$
4	S	5	C	$4\frac{1}{4}$	H	$4\frac{1}{2}$	J	4	D	6
5	H	2	D	$2\frac{1}{2}$	J	1	C	$1\frac{1}{2}$	S	$2\frac{1}{4}$

lowest so gets rank 2, and so on giving the following table of ranks.

Taster	C	D	H	J	S	
1	2	5	4	1	3	
2	3	5	4	1	2	
3	2	5	4	1	3	
4	2	5	3	1	4	
5	2	5	3	1	4	
Total	11	25	18	5	16	75

Then the ranks are totalled for each treatment. Again there is a simple check on the ranking. If there are p treatments and r blocks, calling the treatment rank totals R_i, the grand total

$$\sum R_i = \tfrac{1}{2}rp(p+1) = \tfrac{1}{2} \times 5 \times 5 \times 6 = 75$$

in this case, so our total is right.

To test if all the beef can be said to come from the same population, we calculate

$$S = \sum R_i^2 - \frac{(\sum R_i)^2}{p} = 1351 - 1125 = 226$$

Tables are available for testing S directly when r and p are small; but for larger experiments such as ours we calculate

$$\frac{12S}{rp(p+1)} = \frac{12 \times 226}{5 \times 5 \times 6} = 18\cdot08$$

and refer this to the table of χ^2 (Appendix Table III, p. 228) with $p-1$ degrees of freedom. In the present case we find that our χ^2 exceeds the tabular value for $P \leqslant 0\cdot01$; so, if the null hypothesis that the medians of the p treatments are equal is true, we have witnessed a very unlikely event. Such a test requires, as in most of the non-parametric tests discussed earlier, that the form of the distribution shall be the same for all positions in the table of results.

Three-way classification

Now we should inquire if we can go further and take out more variation by having extra classifications. There are two approaches. First we can look upon the randomized-block design as being a factorial arrangement of blocks as one factor and treatments as the other, and proceed from this to the idea that if we could recognize two sources of natural variation which could be factorially combined with each other and with treatments we could have a three-way classification.

An example occurs with pig experiments, where the main extraneous sources of variation are genetic variation and sex. We can control genetic variation to a very large extent if we use litters of pigs as blocks. Then all the pigs in any block have the same parents and, as a bonus towards controlling variation, were all born on the same day, so we want all treatments represented equally in all litters. Similarly, sex variation can be controlled by having blocks all of one sex and containing all treatments equally. Thus a three-way classification will mean that all treatments must appear equally on each sex in each litter. Many such experiments were made in an experimental piggery especially built for this design in Cambridge in the mid-1930s. In this case there were five litters for each experiment; three treatments were given and each treatment was allocated at random to one of each sex in each litter, giving a lay-out for treatments a_0, a_1 and a_2 such as that shown:

Litters	I		II		III		IV		V	
	♂a_0	♀a_1	♂a_1	♀a_0	♂a_2	♀a_2	♂a_0	♀a_2	♂a_0	♀a_0
	♂a_2	♀a_0	♂a_2	♀a_1	♂a_1	♀a_1	♂a_1	♀a_0	♂a_2	♀a_1
	♂a_1	♀a_2	♂a_0	♀a_2	♂a_0	♀a_0	♂a_2	♀a_1	♂a_1	♀a_2

The skeleton analysis of variance for l litters, s sexes and p treatments and for the present case is

	D.F.		Expected mean square
Litter (L)	$(l-1)$	4	$\sigma^2_{TLS} + p\sigma^2_{LS} + s\sigma^2_{TL} + ps\sigma^2_L$
Sex (S)	$(s-1)$	1	$\sigma^2_{TLS} + p\sigma^2_{LS} + lpS^2$
Litter × Sex ($L \times S$)	$(s-1)(l-1)$	4	$\sigma^2_{TLS} + p\sigma^2_{LS}$
Treatment (T)	$(p-1)$	2	$\sigma^2_{TLS} + s\sigma^2_{TL} + slT^2$
$T \times L$	$(p-1)(l-1)$	8	$\sigma^2_{TLS} + s\sigma^2_{TL}$
$T \times S$	$(p-1)(s-1)$	2	$\sigma^2_{TLS} + l(TS)^2$
$T \times L \times S$	$(p-1)(l-1)(s-1)$	8	σ^2_{TLS}
Total	$lps-1$	29	

The sums of squares can be obtained easily by the methods already discussed.

Error for testing treatments needs careful thought. First we must decide for each factor whether it exhibits a fixed or random effect. In the present case the treatments that we want to test, which might be three different diets, would be a fixed-effect factor. The other two factors have been introduced to remove some of the natural variation, and are often called *control factors* for that reason. If we omitted one of the litters and substituted another, we should still have the same experiment, so it would be reasonable to consider litters as a random-effect factor; the results of our experiment will apply to the population of litters (perhaps one breed) from which our sample of litters was drawn. On the other hand, if we left out one of the sexes, we should have a different experiment and different conclusions, e.g. effects of treatments only on male pigs and not on pigs as they come, more or less equal numbers of males and females; so we should consider sex as a fixed-effect factor. This leads to the expected mean squares given in the skeleton analysis-of-variance table on p. 145. It is then seen that Treatment should be tested with Treatment × Litter as an error and, if Sex or Treatment × Sex require testing, the appropriate errors are Sex × Litter and Treatment × Sex × Litter, respectively.

In fact, if there is but one random-effect factor in an experiment, there is no need to go through the whole model in this way, because it always turns out that the treatments can be tested against their interaction with the random-effect factor. However, in a multiway classification the two-factor interactions usually have few degrees of freedom, and consequently the appropriate error is not well based and will detect only large differences. It is therefore useful to investigate the various error terms to see if any combinations can be made. Thus before testing treatments we could test $T \times L$ (the theoretically appropriate error for treatments) against $T \times L \times S$, which is the appropriate error for $T \times L$. If it was significantly different, then we have no choice but to use $T \times L$ for testing T but, if it was not, then we have no evidence to suggest that σ_{TL}^2 has any value other than zero, so the expected mean square for treatments degrades to $\sigma_{TLS}^2 + slT^2$ and for $T \times L$ to σ_{TLS}^2 or the same as $T \times L \times S$. Then we could combine the S.S. of $T \times L$ and $T \times L \times S$ to give an error term with 16 degrees of freedom in this case for testing the treatments. Likewise we could test $L \times S$ against $T \times L \times S$ and, if they did not differ, combine their sums of squares to produce an error term for S of 12 D.F. instead of the theoretically appropriate one with only 4 D.F. for testing.

It must be remembered that variances are themselves variables and so, to avoid being misled by combining errors simply because the data suggest that it would be a good thing to do, definite null hypotheses with levels of probability for acceptance should be built into the experiment before it begins. In this case we would not necessarily use the conventional 5 per cent level of probability; we would be guided by our biological expectation of whether the variation should be different or not, and use a more stringent test if it is going against previous biological experience; Sokal and Rohlf (1969, p. 266) give a good set of rules for such tests.

One further problem is seen by inspecting the expected mean square column of the skeleton analysis-of-variance table, namely that there is no single mean square appropriate to testing litters. In this case it would not matter, but in an experiment where there is more than one random-effect local control factor the same difficulty can arise for testing treatments. In most cases there will be possible combinations of mean squares which give the necessary expected mean square. In the present case the M.S. for $L \times S$ plus that for $T \times L$ minus that for $T \times L \times S$ is

$$\sigma^2_{TLS} + p\sigma^2_{LS} + \sigma^2_{TLS} + s\sigma^2_{TL} - \sigma^2_{TLS} = \sigma^2_{TLS} + p\sigma^2_{LS} + s\sigma^2_{LT}$$

which would be appropriate for testing litters. It is an important point of principle in biological experimentation to investigate the proposed analysis of variance before starting the experiment, to ensure that appropriate errors are available for testing all the null hypotheses.

Latin square

There is another method of removing two sources of variation in an experiment, and this method is more commonly used. Let us suppose that we are designing a crop experiment which has five treatments, which for the time being we will call A, B, C, D, E. Let us further suppose that the field we have available for this experiment has quite a distinct slope from north to south; and from previous observation of crops it is well known that the fertility of the soil decreases from west to east as shown in figure 8.1.

If we use a randomized-block design, putting our blocks north and south of each other (figure 8.1a) we shall take out all the variation due to the slope in the blocks sum of squares, and all treatments will appear equally at any level of slope, but the variation due to the changes of fertility will be left to increase the within-block variation or error. In the randomization shown, B in the southern block may yield less than B in the

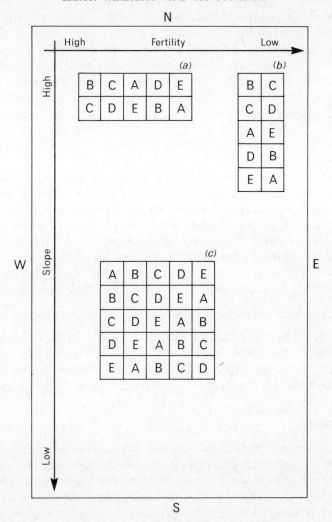

Figure 8.1—Possible field layouts of randomized blocks and a Latin square.

northern block, simply because of its position in the block. Randomization will ensure that this does not bias the treatment effects, but it will mean that the error will be large, and so small differences will not be detected unless we have very considerable replication. Similarly, if we put our blocks east and west of each other (figure 8.1b), we shall remove the fertility variation in the blocks sum of squares, but be left with the slope variation included in the treatment effects and error. What we need is to be able to put the blocks both ways at the same time. This we can do if we have the same number of replications as we have treatments (figure 8.1c).

Such a design is called a *Latin Square*, the main features being that each letter (which means each treatment in an experiment) appears once and once only in each row, and once and once only in each column; and there are therefore as many rows as there are letters or treatments, and as many columns as there are letters or treatments. Conventionally we speak of an $r \times r$ Latin square where r tells us how many rows, columns and letters we are talking about, in this case 5×5. The skeleton analysis of variance is:

	D.F.
Rows	$r-1$
Columns	$r-1$
Treatments	$r-1$
Error	$(r-1)(r-2)$
Total	r^2-1

We see that we remove from the total variation any which is associated with rows and any associated with columns, and consequently the error variation will be smaller. In our problem the variation between rows would be that due to slope, and that between columns would be variation due to fertility changes. Many comparisons of Latin squares and randomized blocks with the same replication have been made; they almost all show that Latin squares lead to lower coefficients of variation than randomized blocks. This is not surprising, since either the rows or the columns will remove the same variation as blocks in a randomized-block design, and any variation removed by the other must reduce the error variation.

Let us suppose that we wished to compare the effects of five different types of phosphatic fertilizer on the yield of potatoes. We have five treatments which we will designate p_1, p_2, p_3, p_4 and p_5, and suppose that we have decided that five replications would be sufficient to test our null hypotheses. So with five treatments and five replications, and the field specified in our figure 8.1, we decide to use a Latin square. The first thing we must do is consider the possibility of randomization in choosing the square for the experiment—there are obviously several possible squares. Following our arguments for one-way classification and randomized blocks, all treatments must have the same chance of being applied to any particular item. This is the rule that we should adopt here, but we would find difficulty in using exactly the same process as in a randomized-block design because, after we had chosen the order of the first row by random numbers, we have an additional restriction in the second row in that no letter must appear in the same column as it did in the first row, and so on. A simpler method is to be able to define all the possible complete squares, and then choose one of them at random.

This definition of the squares is fairly easily done by setting out all the possible squares with letters in natural order in the first row and first column—these are usually referred to as *standard squares*. Then all possible squares can be obtained by all possible permutations of the rows and columns. So if we choose one of the standard squares at random, and arrange the rows and the columns at random, every square has had the same opportunity of being chosen. All standard squares up to 9×9 are set out in Fisher and Yates (1963); with the aid of such tables it is a simple operation. Without such tables it is simple to write down all the standard squares for 3×3 (there is only one) and 4×4 (there are four) and randomize the rows and columns. For higher squares there are many more possible standard squares and it becomes a very tedious process; fortunately, because of the large number available, it is considered reasonable to approximate. The method commonly used is this:

Write down any square of the right size (the most obvious is the one in figure 8.1); then do not two but *three* randomizations, randomize the rows, randomize the columns, and randomize the treatments to the letters. To do this in the present case we want three sets of random digits from 1 to 5.

Rows	2	3	4	5	1
Columns	4	1	2	3	5
Treatments	4	3	5	1	2

Numbering the rows of the original square in natural order from the top, we rearrange them in the order given by the random digits for rows.

```
1  A B C D E           B C D E A
2  B C D E A           C D E A B
3  C D E A B  becomes  D E A B C
4  D E A B C           E A B C D
5  E A B C D           A B C D E
```

Now numbering the columns of this second square in natural order from left to right and rearranging them in the order given by the second set of random digits,

```
1 2 3 4 5
B C D E A              E B C D A
C D E A B              A C D E B
D E A B C   becomes    B D E A C
E A B C D              C E A B D
A B C D E              D A B C E
```

ERROR VARIATION AND ITS CONTROL

Finally considering the third set of random digits to be the levels of p we allocate them to the letters set out in natural order.

$$\begin{array}{ccccc} A & B & C & D & E \\ p_4 & p_3 & p_5 & p_1 & p_2 \end{array}$$

and

$$\begin{array}{ccccc} E & B & C & D & A \\ A & C & D & E & B \\ B & D & E & A & C \\ C & E & A & B & D \\ D & A & B & C & E \end{array} \quad \text{becomes} \quad \begin{array}{ccccc} p_2 & p_3 & p_5 & p_1 & p_4 \\ p_4 & p_5 & p_1 & p_2 & p_3 \\ p_3 & p_1 & p_2 & p_4 & p_5 \\ p_5 & p_2 & p_4 & p_3 & p_1 \\ p_1 & p_4 & p_3 & p_5 & p_2 \end{array}$$

When the data have been obtained they are set out on a plan as in Table 8.5.

Table 8.5—Results and analysis of a Latin square

Layout and yields

					Total (R)
p_2 62·3	p_3 61·3	p_5 62·5	p_1 63·8	p_4 75·0	324·9
p_4 64·1	p_5 68·4	p_1 62·9	p_2 66·2	p_3 77·4	339·0
p_3 69·2	p_1 55·8	p_2 67·8	p_4 71·3	p_5 74·8	338·9
p_5 65·0	p_2 68·7	p_4 69·8	p_3 76·0	p_1 70·9	350·4
p_1 63·3	p_4 75·0	p_3 69·3	p_5 78·0	p_2 75·4	361·0
Totals (C) 323·9	329·2	332·3	355·3	373·5	1714·2 (G)

	p_1	p_2	p_3	p_4	p_5
Treatment totals (T)	316·7	340·4	353·2	355·2	348·7

Analysis of variance

	D.F.	S.S.	M.S.	F
Rows	4	147·81	36·95	
Columns	4	350·23	87·56	
Treatments	4	196·73	49·18	4·74
Error	12	124·54	10·38	
Total	24	819·31		

We then add across to obtain the row totals, add downwards to obtain the column totals, and sum the yields of p_1s, p_2s, etc., to give the treatment totals. Then we can work out the sums of squares for the analysis of variance table as:

Rows, $\dfrac{\sum R^2}{r} - \dfrac{G^2}{r^2} = \dfrac{324 \cdot 9^2 + 339 \cdot 0^2 + 338 \cdot 9^2 + 350 \cdot 4^2 + 361 \cdot 0^2}{5} - \dfrac{1714 \cdot 2^2}{25}$

$= 117\,687 \cdot 08 - 117\,539 \cdot 27 = 147 \cdot 81$

Columns, $\dfrac{\sum C^2}{r} - \dfrac{G^2}{r^2} = \dfrac{323 \cdot 9^2 + 329 \cdot 2^2 + 332 \cdot 3^2 + 355 \cdot 3^2 + 373 \cdot 5^2}{5} - \dfrac{1714 \cdot 2^2}{25}$

$= 117\,889 \cdot 50 - 117\,539 \cdot 27 = 350 \cdot 23$

Treatments, $\dfrac{\sum T^2}{r} - \dfrac{G^2}{r^2} = \dfrac{316 \cdot 7^2 + 340 \cdot 4^2 + 353 \cdot 2^2 + 355 \cdot 2^2 + 348 \cdot 7^2}{5} - \dfrac{1714 \cdot 2^2}{25}$

$= 117\,736 \cdot 00 - 117\,539 \cdot 27 = 196 \cdot 73$

Total, $\sum x^2 - \dfrac{G^2}{r^2} = 62 \cdot 3^2 + 61 \cdot 3^2 + \ldots + 75 \cdot 4^2 - \dfrac{1714 \cdot 2^2}{25}$

$= 118\,358 \cdot 58 - 117\,539 \cdot 27 = 819 \cdot 31$

The error S.S. is obtained by difference and, although its mean square represents a mixture of interactions, it is appropriate for testing the treatments, provided rows and columns represent random effects and provided the usual assumptions for an F-test hold. Calculating the mean squares in the usual way shows the advantage of using a Latin square in this particular instance. Both rows and columns are associated with much more variation than the error and, had we used a randomized-block design even with the better of the two orientations (i.e. with blocks as columns of our square) we should have had a much higher error mean square, and consequently should not have been able to detect such small treatment differences.

A Latin-square design can be used whenever there are two distinct sources of variation to be controlled, e.g. in a supermarket it was desired to find out if different sorts of pack affected the sales of soap. There were two obvious sources of variation: day of the week, since housewives might, on average, buy different quantities on different days; and position in the store, some buy the first thing they see. There were 6 packs to be compared, 6 suitable counters in the store, and 6 days of the shopping week; so a 6×6 Latin square was used with counters as rows and days as columns. All six packs appeared every day, and each pack appeared on each counter once, and once only, during the week.

In experiments with animals it is often possible to use a Latin square where individual animals form the columns, and the occasions of receiving the treatment form the rows. A more sophisticated arrangement of this type is shown on pp. 201–207, but it should be said here that care is needed to ensure that the assumptions of the model are met. In particular, in such an experiment each animal will be progressively older as the occasions proceed, and one of the assumptions is that the treatment effect is the same from animal to animal, and from occasion to occasion. If the experiment goes on for a long time, it would be unreasonable to believe that a particular feeding treatment, say, would produce the same live-weight increase in a young animal as it would in an older one; then a Latin square should not be used, for wrong conclusions may be drawn.

The Latin square is also useful when we know that we have variation of a linear nature, but do not know which direction it takes. An example is in research with air-borne diseases, when there might be sources of inocculum all round a field, and where plants which get infected first might suffer most. If we used a randomized-block design, it would be a matter of chance whether we had the right orientation for any particular attack. For example, with blocks arranged as follows:

```
    A   B   C        N
    A   C   B        ↑
    B   A   C
```

an attack from the north would affect all treatments equally, but one from the west would affect A seriously, B moderately and C very little, whereas with the Latin square we get all treatments affected equally no matter which direction the pathogen comes from.

```
    A   B   C
    B   C   A
    C   A   B
```

Only if the effect is non-linear can treatments receive different intensities, no matter what randomization of the squares is used.

There are, however, limitations to the use of Latin squares. When we have more than 6 treatments we often find that the Latin square demands more replication than we require for testing our hypotheses, and consequently we should have to do more work than necessary. On the other hand, if we have only 3 or 4 treatments, a single Latin square has insufficient replication for many purposes, but in this case we can get more precision by replicating the square. For example, if we had two substances alleged to control eel-worms, we might require three treatments, s_0 (a control), s_1 and s_2 and, knowing little of the distribution of eel-worms in the soil, might prefer a Latin square, although to get significance for the sort of differences we expect might require 9 replications. Here we would set out side by side three independently randomized squares.

```
s₀  s₂  s₁      s₂  s₁  s₀      s₂  s₀  s₁
s₂  s₁  s₀      s₀  s₂  s₁      s₀  s₁  s₂
s₁  s₀  s₂      s₁  s₀  s₂      s₁  s₂  s₀
```

The analysis of variance would be:

	D.F.
Squares	2
Rows within squares	6
Columns within squares	6
Treatments	2
Error	10
Total	26

The calculation of the sums of squares follows the usual pattern. We could set out the yields on a plan and obtain totals for each row (R), column (C), square (S), treatment (T) and Grand total (G). Then S.S. for squares

$$= \frac{S_1^2 + S_2^2 + S_3^2}{9} - \frac{G^2}{27}$$

S.S. for rows within squares

$$= \frac{R_1^2 + R_2^2 + R_3^2 + \ldots + R_9^2}{3} - \frac{S_1^2 + S_2^2 + S_3^2}{9}$$

Likewise S.S. for columns within squares

$$= \frac{\sum C^2}{3} - \frac{\sum S^2}{9}$$

For treatments the S.S.

$$= \frac{\sum T^2}{9} - \frac{G^2}{27}$$

The total sum of squares is obtained in the usual way and error S.S. obtained by difference.

CHAPTER NINE

CONFOUNDING

When randomized-block or Latin-square designs are used, the experimental material is divided into more homogeneous groups, i.e. blocks or rows or columns, and treatments are all represented equally in each group, so that variation between groups can be removed from the treatment differences and from the error variation. It will be obvious that success in achieving a low error mean square will depend on how homogeneous the individual groups can be made, and from our knowledge of natural variation it will be obvious that the bigger the group the less homogeneous will it be. The larger the number of treatments, the larger must be the group or block, if it is to accommodate one representative of each treatment; so blocks, rows, and columns all become less efficient in controlling extraneous variation as the number of treatments increases. There are three sorts of design to get over this problem.

Split-plot experiments

The first is known as a *split-plot design*. To use this the treatments need to be considered as made up of the combinations of at least two factors. As an example, Table 9.1 shows an experiment in which there are combinations of four varieties (V) and two levels of application of nitrogen (N) in a field experiment.

We have chosen to have five replications, so start as with a randomized-block design by dividing the area of land allocated to the experiment into five blocks. We then divide each block into four plots and allocate the four varieties to these plots, which we shall call main plots, by the process of randomization. Next we divide each main plot into two sub-plots, and allocate the two nitrogen treatments to each pair of sub-plots by separate and independent randomizations. Using random digits the five sets of four

Table 9.1—The layout and analysis of a split-plot design

Layout and yields (cwt/ac)

Block I	v_2		v_1		v_0		v_3	
	n_1	n_0	n_1	n_0	n_1	n_0	n_0	n_1
	47·3	33·4	42·4	37·9	39·8	38·0	41·7	43·0

Block II	v_3		v_2		v_1		v_0	
	n_1	n_0	n_0	n_1	n_0	n_1	n_1	n_0
	41·2	45·1	44·9	44·3	33·2	40·3	45·1	45·8

Block III	v_0		v_1		v_2		v_3	
	n_0	n_1	n_0	n_1	n_1	n_0	n_0	n_1
	28·0	37·1	46·1	52·6	45·2	45·3	44·2	44·7

Block IV	v_2		v_3		v_1		v_0	
	n_1	n_0	n_1	n_0	n_1	n_0	n_0	n_1
	50·4	35·8	50·1	50·5	49·4	39·1	47·0	44·6

Block V	v_0		v_1		v_3		v_2	
	n_1	n_0	n_0	n_1	n_1	n_0	n_1	n_0
	39·4	43·1	43·3	49·5	43·4	41·8	41·3	44·3

Main plot totals (M)   Block totals (B)

80·7	80·3	77·8	84·7		323·5
86·3	89·2	73·5	90·9		339·9
65·1	98·7	90·5	88·9		343·2
86·2	100·6	88·5	91·6		366·9
82·5	92·8	85·2	85·6		346·1

Grand total (G)   1719·6

Treatment totals (T)

	v_0	v_1	v_2	v_3	Total (N)
n_0	201·9	199·6	203·7	223·3	828·5
n_1	206·0	234·2	228·5	222·4	891·1
Total (V)	407·9	433·8	432·2	445·7	1719·6

Analysis of variance

Sources of variation	D.F.	S.S.	M.S.	F
Blocks	4	120·81		
Varieties	3	75·41	25·14	<1
Error (a)	12	432·26	36·02	
Main plots	19	628·48		
Nitrogen	1	97·97	97·97	7·24*
Varieties × nitrogen	3	85·03	28·34	2·09
Error (b)	16	216·52	13·53	
Total	39	1028·00		

*$P < 0.05$

numbers for varieties were:

$$\begin{array}{cccc} 2, & 1, & 0, & 3 \\ 3, & 2, & 1, & 0 \\ 0, & 1, & 2, & 3 \\ 2, & 3, & 1, & 0 \\ 0, & 1, & 3, & 2 \end{array}$$

which gave the arrangement of varieties and then, by writing n_0 first for even numbers from the random number table and n_1 first for odd numbers, the arrangement of the nitrogen treatments was obtained from

1 1 1 0 1 0 0 1 0 0 1 0 1 1 1 0 1 0 1 1

If there are r blocks, p varieties, and q levels of nitrogen, the skeleton analysis of variance is

	D.F.
Blocks	$(r-1)$
Variety	$(p-1)$
Error (a)	$(r-1)(p-1)$
Main plots	$(rp-1)$
Nitrogen	$(q-1)$
Variety × nitrogen	$(p-1)(q-1)$
Error (b)	$p(r-1)(q-1)$
Total	$rpq-1$

This is most easily built up by considering the field operations. First the field was divided into r blocks of land which must have $(r-1)$ D.F. Then each block was divided into p main plots to which varieties were allocated, one of each to each block, so varieties with $(p-1)$ D.F. are orthogonal to blocks; we can immediately write down their interaction with $(r-1)(p-1)$ D.F. and, since this is an interaction with a random-effect factor (blocks), we can call it an *error term*. This is all the information we can get from the main plots, so the degrees of freedom should sum to one less than the number of main plots $(rp-1)$, which they do. The next field procedure was to "split" each main plot into q sub-plots and allocate these to q levels of nitrogen application. Nitrogen application with $(q-1)$ D.F. is therefore orthogonal to main plots and, as in any two-factor arrangement, there will be an interaction between nitrogen and main plots with $(rp-1)(q-1)$ D.F. However, since the main plot variation can be split into three components; blocks, varieties, and block × variety interaction, the interaction between main plots and nitrogen application can also be divided into three components, namely, blocks × nitrogen application, variety × nitrogen application, and blocks × variety × nitrogen. Of these, varieties × nitrogen with

$(p-1)(q-1)$ D.F. is an interaction between two fixed-effect factors that we shall wish to test, but the other two are interactions with the random-effect factor and can be combined to form another error term with

$$(r-1)(q-1)+(r-1)(p-1)(q-1) = p(r-1)(q-1) \text{ D.F.}$$

The D.F. column should then be checked by ensuring that it sums to one less than the total number of sub-plots.

In technical terms, the varieties are known as *main treatments* because they occupy the main plots (not because they are the most important treatments—they usually are not) and the levels of nitrogen are called the *sub-treatments* because they occupy the sub-plots. Error (*a*) is the main-plot error and is appropriate for testing main treatments, whilst error (*b*) is the sub-plot error and is appropriate for testing sub-treatments and the interactions between sub-treatments and main treatments. Looking at the layout in Table 9.1 it is seen that error (*a*) is made up of variation within whole blocks, i.e. variation within large groups, in this case of 8 plots, whereas error (*b*) is made up of variation within small groups, namely main plots which are only a quarter the size of a block. Therefore it is logical to suggest that error (*b*) will be smaller than error (*a*). This has been tested many times with field experiments and is found to be so. This design is useful when there are at least two factors and we want to test some of the factors with a high degree of precision, but are prepared to test one or more of the factors with less precision than the others. This is often the case when extra factors are put into an experiment simply to test the hypotheses over a wider range of conditions. Then we are not at all interested in a comparison of the "condition" treatments, but only in our original treatments and their interactions with the conditions.

To calculate the sums of squares in Table 9.1 we first need to add together yields of sub-plots in each main plot to get main plot totals (M) from which we can get block totals (B). The treatment combination totals (T) are best set out in a two-way table with main treatments along one margin and sub-treatments along the other. We can then add the rows to give, in our case, the total yields of the nitrogen treatments (N) and add the columns to give the variety totals (V). Adding these treatment totals gives us the grand total (G), and we can check that we have the same grand total by adding the blocks.

The sum of squares for blocks is

$$\frac{\sum B^2}{pq} - \frac{G^2}{rpq} = \frac{323 \cdot 5^2 + \ldots + 346 \cdot 1^2}{8} - \frac{1719 \cdot 6^2}{40} = 120 \cdot 81$$

S.S. for varieties $= \dfrac{\sum V^2}{rq} - \dfrac{G^2}{rpq} = \dfrac{407 \cdot 9^2 + \ldots + 445 \cdot 7^2}{5 \times 2} - \dfrac{1719 \cdot 6^2}{40}$

$= 75 \cdot 41$

S.S. for main plots $= \dfrac{\sum M^2}{q} - \dfrac{G^2}{rpq} = \dfrac{80 \cdot 7^2 + 80 \cdot 3^2 + \ldots + 85 \cdot 6^2}{2} - \dfrac{1719 \cdot 6^2}{40}$

$= 628 \cdot 48$

and error (a) is got by difference, i.e.

S.S. of main plots $-$ S.S. of blocks $-$ S.S. of varieties $= 432 \cdot 25$

S.S. for nitrogen can be got by the difference method as

$$\dfrac{(N_1 - N_0)^2}{2rp} = \dfrac{62 \cdot 6^2}{40} = 97 \cdot 97$$

S.S. for varieties $\times$ nitrogen can be got by the general method as:

$\dfrac{\sum T^2}{r} - \dfrac{G^2}{rpq} -$ S.S. for varieties $-$ S.S. for nitrogen $= \dfrac{201 \cdot 9^2 + \ldots + 222 \cdot 4^2}{5}$

$- \dfrac{1719 \cdot 6^2}{40} - 75 \cdot 41 - 97 \cdot 97 = 85 \cdot 03$

In the special case of only two sub-treatments it could also be got from the differences between n_1 and n_0 for each variety.

Total S.S. is got in the usual way and turns out to be $1028 \cdot 00$.

S.S. for error (b) is then Total S.S. $-$ Main plots S.S. $-$ Nitrogen S.S. $-$ Varieties $\times$ Nitrogen S.S. $= 216 \cdot 52$.

Again in the special case of only two sub-treatments, S.S. for error (b) could be got directly from the differences between n_1 and n_0 for each main plot.

Now main treatments can be tested by means of an F-test using error (a), i.e.

$$F_{(3,12)} = \dfrac{\text{Variety M.S.}}{\text{Error }(a)\text{ M.S.}} = \dfrac{25 \cdot 14}{36 \cdot 02} = < 1$$

and sub-treatments and sub-treatments $\times$ main treatments interaction using error (b)

$$F_{(1,16)} = \dfrac{\text{Nitrogen M.S.}}{\text{Error }(b)\text{ M.S.}} = \dfrac{97 \cdot 97}{13 \cdot 53} = 7 \cdot 24$$

and

$$F_{(3,16)} = \frac{\text{Varieties} \times \text{Nitrogen M.S.}}{\text{Error }(b)\text{ M.S.}} = 2\cdot09$$

Both main and sub-treatments and their interaction can be sub-divided in any of the orthogonal ways shown in Chapters 5, 6 and 7, and each sub-division tested in exactly the same way as with any other design.

If t-tests or confidence limits are required, standard errors can be supplied, though we have to think carefully which error is appropriate. Table 9.2 shows all possible standard errors for a table of treatment means. This table is simply obtained from the treatment totals in Table 9.1 by dividing each cell in that table by 5, the number of replications, the right-hand marginal totals by 20 and the bottom marginal totals by 10.

Table 9.2—Treatment means and standard errors

	v_0	v_1	v_2	v_3	Mean	S.E.
n_0	40·4	39·9	40·7	44·7	41·4	$\pm 0\cdot82$
n_1	41·2	46·8	45·7	44·5	44·6	
Mean	40·8	43·4	43·2	44·6		
S.E.		$\pm 1\cdot90$				

S.E. for comparing two levels of nitrogen on the same variety = $\pm 1\cdot64$
S.E. for comparing two varieties at the same or at different levels of nitrogen = $\pm 2\cdot23$

Looking first at the margins, the comparisons of N means are simply comparisons of figures obtained from sub-plots in that every n_0 has a corresponding n_1 in the same main plot, so any variation attached to the difference between them must be the sub-plot type of variation. Therefore the appropriate S.E. of a nitrogen mean (in general a sub-treatment mean) is

$$\sqrt{\frac{E_b}{rp}} = \sqrt{\frac{13\cdot53}{5 \times 4}} = 0\cdot82$$

At the other margin we have variety or main treatment means and in this case differences between sub-plots do not enter in at all. We have simply the sum of two sub-plots to give us each main plot total and any variation between main treatment means is due to variation between main plots; hence the S.E. for a variety mean or main treatment mean is

$$\sqrt{\frac{E_a}{rq}} = \sqrt{\frac{36\cdot02}{5 \times 2}} = 1\cdot90$$

Now suppose we wished to compare the two nitrogen levels on one particular variety, say v_0n_0 and v_0n_1. Whenever v_0n_0 occurs, v_0n_1 appears in the same main plot, so that any natural variation between them is the variation due to sub-plots; thus their S.E. is based on error (b) and must be

$$\sqrt{\frac{E_b}{r}} = \sqrt{\frac{13\cdot53}{5}} = 1\cdot64$$

But suppose we want to compare v_0n_0 with v_1n_0 or with v_1n_1. Then in every comparison the two treatments would not only occupy different sub-plots, but would also be in different main plots, as can be seen from the layout in Table 9.1, so the comparison will be influenced by both error (a) and error (b), and these have to be combined in an appropriate proportion to provide a single value for the standard error. In fact the appropriate proportion simply depends upon the degree of sub-division of the main plots and the correct error variance is

$$\frac{E_a + (q-1)E_b}{q}$$

so the S.E. of a VN mean for this comparison would be

$$\sqrt{\frac{E_a + (q-1)E_b}{rq}} = \sqrt{\frac{36\cdot02 + 13\cdot53}{5 \times 2}} = 2\cdot23$$

If a t-test is required in any of these cases we must use the correct number of degrees of freedom when looking up t. In the case of main-treatment differences where only error (a) is involved, t has the degrees of freedom of error (a). Likewise where only error (b) is concerned, t has the degrees of freedom of error (b), but in the last case, where both errors are concerned, there is no true t and we have to calculate an approximate weighted average of t. If t_a = the tabular value for the D.F. of E_a and t_b, the corresponding value for E_b, then an approximate value is

$$t = \frac{(q-1)E_b t_b + E_a t_a}{(q-1)E_b + E_a} = \frac{13\cdot53 \times 2\cdot12 + 36\cdot02 \times 2\cdot18}{13\cdot53 + 36\cdot02} = 2\cdot16$$

in the present case. This value of t is biased upwards, so is safe in that we get too few significant results. Thus, if the null hypothesis required us to assess the difference between v_0 and v_3 in the absence of nitrogen, we should require a difference of $2\cdot23 \times \sqrt{2} \times 2\cdot16 = 6\cdot81$ for significance at the 5 per cent level, or $\pm 6\cdot81$ would give the 95 per cent confidence interval of the difference.

The S.E.s in the table show that much smaller true differences can be detected between the sub-treatment (n-levels) means than between the main-treatment or variety means and that, if there are a number of null hypotheses involved in the interaction, those which involve comparing two sub-treatments at the same level of the main treatment will be tested with greater precision than others. In the present case we would be 95 per cent confident that v_1 and v_2 respond to nitrogen, but we could not be 95 per cent confident that the varieties themselves were different at either level of nitrogen. Thus, if this design is used to obtain greater precision for some factors or interactions, great care must be taken to ensure that the tests of these factors, and the tests within the interaction, are included in the design in such a way that they are made with the sub-plot error. Of course, as with a randomized-block design, the sub-plot error will be reduced only if the variation between groups (in this case main-plots) is greater than the variation within the groups, and less precision would be obtained if the groups were simply made up at random, because the degrees of freedom for each error would obviously be less than the degrees of freedom for the combined error if a randomized-block design was used. Thus, using a split-plot design, simply because there is a factorial arrangement of treatments, is of no help in reducing error unless there are some biological circumstances which cause different amounts of variation. This is sometimes the case in field experiments where the soil is not very uniform. With plants amenable to vegetative propagation, variation between clones may be greater than variation between plants derived from the same clone. When considering parts of plants, variation between two leaves on the same plant may be less than the variation between leaves on different plants, so plants could form main plots and leaves sub-plots. Likewise with animals, whole litters might form main plots and individual members of the litters receive the sub-treatments; then the sub-plot error would usually be smaller than the main plot error, because it would contain less genetic variation or variation due to age of animal. Experiments on certain organs of animals can often be treated in the same way with animals as main plots.

However, split-plot designs are sometimes used for convenience rather than to reduce error for some particular factor. This occurs in two ways.

(1) One factor may require a large quantity of material if it is to be applied realistically, whilst another factor can be applied to small quantities. A one-way classification or randomized-block design would require a large quantity of material for each replication of each combination of the two factors, whereas the split-plot design allows the factor requiring the large quantity to be applied as a main treatment, and the material can then be "split" to receive the levels of the other factor.

(2) In an experiment primarily designed to test a factor which requires a large amount of material for each replicate, we may make more economical use of the material by combining this experiment with another involving a factor requiring only a small quantity of material for each replication of each level; in such a case the interaction may be of no interest at all.

Work with trees often suggests this kind of design in that cultural treatments, such as mulching, irrigating, fertilizing, and spraying chemicals to control pests or diseases, all require an area of trees with guard rows round each plot. Thus we can be sure that the trees being measured really received the full treatment, but treatments such as varieties or pruning can often be applied to single trees. If we wished to see if four particular varieties differed in their response to spraying with three particular chemicals, we could use a split-plot design with chemicals as main treatments applied to main plots made up of one tree of each of the four varieties planted in the form of a square and completely surrounded and protected by 12 other trees. A complete replicate would require 48 trees on that basis, but if each variety × chemical plot were to be sprayed separately, to get the same protection for a single tree it must be surrounded by eight others, so 108 trees per replicate would be required.

When using split-plot designs for convenience in this way, great care must be taken to ensure that all factors are tested with reasonable precision. This often means having more replication than would be required with a simpler design, so that there are sufficient degrees of freedom for error (a) to give reasonable tests. For example if we had three factors (A, B and C) each at two levels and decided to use a split-plot design with four replicates (which might be perfectly adequate in a randomized-block design) putting A into main plots because it required large amounts of material, the analysis of variance would be

Source of variation	D.F.
Blocks	3
A	1
Error (a)	3
Main plots	7
B	1
AB	1
C	1
AC	1
BC	1
ABC	1
Error (b)	18
Total	31

with a very poor test of A based on an error with only 3 D.F. To get a reasonable test for A we should need something like 12-fold replication (giving 11 D.F. for error (a)). This would mean getting three times as many observations as required in a randomized-block design, so much of the convenience disappears. If on the other hand we compromised and allocated all combinations of A and B to main plots, the analysis of variance would be

Source of variation	D.F.
Blocks	3
A	1
B	1
AB	1
Error (a)	9
Main plots	15
C	1
AC	1
BC	1
ABC	1
Error (b)	12
Total	31

We have a much better test for main treatments, and one more replicate would give a perfectly satisfactory experiment compared with a randomized-block design. In general, using a split-plot design for convenience is admirable if the main treatment factors have several levels, but if they have few levels the extra replication required to test them adequately often means that no advantage is obtained.

Another possible use of split-plot designs is in modifying long-term experiments. Experiments with trees or ecological studies often have to be carried on for many years before the looked-for data are obtained but, during the course of the experiment, new knowledge may be obtained which suggests other null hypotheses that should be tested. To set up a new experiment might again require a large number of preliminary years, so it would be much better to be able to incorporate the new treatments into the present experiment. Provided each original "plot" is capable of being split, the new factor can be introduced as a sub-treatment. It should be realized that this will not affect the precision of the tests of the original factors, although the new factor and its interactions with the old factors will most likely be tested with greater precision. Also the amount of work will be increased, in fact multiplied by the number of levels in the new factor. Thus we need to be very sure of the value of the extra information before embarking on such an exercise.

There is sometimes another way of introducing the new factor without increasing the number of observations. For example, if we had a long-term experiment with eight treatments in four randomized blocks, and wished to introduce a new factor at two levels, we could allocate two blocks at random to each level of the new factor giving an analysis of variance:

Source of variation	D.F.
New factor	1
Error (a)	2
Blocks	3
Old treatments	7
New × old	7
Error (b)	14
Total	31

Now the new factor *per se* is tested with very little precision, but its interaction with the old treatments has a satisfactory test, and often the reason for introducing the new factor is to test such interaction. In this case the degrees of freedom for the error testing the old treatments are reduced, but usually this reduction is not serious unless there was originally very little replication, or the new factor has many levels, when, of course, this device should not be used.

Finally it should be stressed that the analysis of a split-plot design assumes that the sub-plot values are independent, so this analysis applies only when the levels of the sub-treatment factors are randomized independently to the sub-plots. A common mistake is to consider that any experiment in which the experimental material is divided up in stages should have a split-plot analysis. For example, experiments with grasses often involve cutting and weighing the foliage several times, so the data can be set out in the form of a two-way table with experimental treatments along one margin and harvests along the other, with all the appearance of a split-plot design in that experimental treatments are obviously involved in variation between plants and the harvests with variation due to time within plants. However, we cannot randomize the harvests; the first must come before the second, whatever happens. In such a design all analyses of variance must be in the form appropriate to the experimental treatments. We can analyse each harvest separately or various combinations of harvest, e.g. the sum of all harvests tells us the effects on total production, or fitting orthogonal polynomials to the harvests of each plot and using the $\sum \xi' y / (r \sum \xi'^2)$ as the variates in analysis of variance will tell us if treatments affect the relationship between yield and harvest.

Confounding interactions with blocks

Another method of using smaller groups of material and reducing the error is known as *confounding*. In fact it is very similar in principle to split-plots but here, instead of applying certain factors to main plots, we apply certain interactions to main plots, so these interactions are tested with little precision or sometimes not tested at all, whilst all the main effects of the factors and the remaining interactions can be tested with greater precision than would be possible with a straightforward randomized-block design. As an example, we might have treatments of all combinations of presence and absence of three drugs (A, B and C). Suppose we are dealing with small animals which normally have only four or so young per litter, so we could get homogeneous groups of four (all full brothers and sisters) but could not possibly get very homogeneous groups of eight. Yet we have eight treatment combinations. We might well think that the second-order interaction, ABC, was of no interest, so we might be prepared to lose all the information about this interaction if we could use more homogeneous groups. In other words, we would be prepared to confound the ABC-effect with groups or confuse it with groups to achieve this.

The contrasts of the various effects in a 2^3 experiment (given on p. 90) show that the ABC-interaction has these coefficients:

	(1)	a	b	ab	c	ac	bc	abc
ABC	−1	+1	+1	−1	+1	−1	−1	+1

We arrived at this by writing the coefficients for each factor separately as $+1$ for presence and -1 for absence, and then multiplying these coefficients together. A little thought on this will show that there is a simpler method for getting any one effect, whether it be main effect or interaction. It is that we put one sign for any combination containing an even number of the factor letters under consideration, and the other sign for those containing an odd number (for this purpose none of the factor letters is taken as an even number). We have put -1 for even numbers (1), ab, ac, bc, i.e. those containing none or two of the letters, and $+1$ for odd numbers a, b, c, abc, i.e. those containing one or three. This method is perfectly general, but by convention we put $+1$ for the set which contains all the factor letters together. Having derived the desired contrast, if we put all combinations with the negative coefficients into one group, or litter in our case, and all those with the positive coefficients in another, this interaction effect cannot be distinguished from the group effect, or is confounded with groups because the measure of the interaction effect is also the measure of any difference between two groups. In our example, if we want four complete replicates, we would take eight complete litters of four animals each and allocate these at random to the + and − sides of our interaction, four litters receiving a, b, c and abc, whilst four receive ab, ac, bc and (1). Then these individual combinations of treatments would be allocated at random to the four animals in each litter. The full analysis of variance would be:

	D.F.
ABC	1
Between litters within ABC	6
Litters	7
A	1
B	1
AB	1
C	1
AC	1
BC	1
Error	18
Total	31

In this case we could in fact test the ABC-interaction using the "between litters" mean square, but it would be a very poor test with an error based on only 6 D.F. In most cases we are not interested in the confounded interaction, so would start the analysis of variance at litters whose S.S. would be simply

$$\frac{\sum L^2}{4} - \text{C.F.}$$

The other sums of squares are obtained in the usual way, and the tests and presentation of the results would be exactly the same as with any 2^n experiment, except that no effect of ABC would be shown, and comparisons involving the effect ABC would not be made, i.e. two-way tables would be in order but a three-way table of treatment means would not be free of litter effects.

Partial confounding

Instead of confounding the ABC-interaction in all replicates, we could confound different interactions in the different replicates, e.g. confound ABC in the first, AB in the second, AC in the third, and BC in the fourth. The appropriate combinations for each block are again worked out from the + and − table, and we should have the sets shown in Table 9.3, with the first replicate divided into a pair of blocks, one contain-

Table 9.3—Layout and analysis of a 2^3 experiment with partial confounding

Layout and yields

Replicate	I		II		III		IV	
Interaction confounded	ABC		AB		AC		BC	
Block	I(i)	I(ii)	II(i)	II(ii)	III(i)	III(ii)	IV(i)	IV(ii)
	(1) 3·27	a 3·96	a 4·10	(1) 4·08	a 4·22	(1) 4·29	b 4·35	(1) 4·57
	ab 3·63	b 3·21	b 3·75	c 4·12	c 4·36	b 3·56	c 4·66	a 4·83
	ac 3·67	c 3·64	ac 4·21	ab 4·72	ab 4·12	ac 3·80	ab 4·19	bc 3·90
	bc 3·15	abc 4·06	bc 3·71	abc 4·74	bc 4·25	abc 4·34	ac 3·92	abc 4·89
Totals (L)	13·72	14·87	15·77	17·66	16·95	15·99	17·12	18·19

Treatment totals and effects

	Treatment totals	Sums and differences				Effects Total	Mean
(1)	16·21	33·32	64·85	130·27	Total	130·27	4·071
a	17·11	31·53	65·42	+4·53	A	+4·53	+0·283
b	14·87	32·38	+2·69	−1·13	B	−1·13	−0·071
ab	16·66	33·04	+1·84	+5·09	AB'	+3·20	+0·267
c	16·78	+0·90	−1·79	+0·57	C	+0·57	+0·036
ac	15·60	+1·79	+0·66	−0·85	AC'	+0·11	+0·009
bc	15·01	−1·18	+0·89	+2·45	BC'	+1·38	+0·115
abc	18·03	+3·02	+4·20	+3·31	ABC'	+2·16	+0·180

C.F. = 530·3210

Analysis of variance

	D.F.	S.S.	M.S.
Blocks	7	3·8982	
A	1	0·6413	
B	1	0·0399	
AB'	1	0·4267	
C	1	0·0102	
AC'	1	0·0005	
BC'	1	0·0794	
ABC'	1	0·1944	
Error	17	1·2007	0·0706
Total	31	6·4913	

ing a, b, c and abc, and the other (1), ab, ac and bc. The second replicate confounding AB would have (1), c, ab and abc (none or two of the letters AB) in one block and a, b, ac and bc (one of the letters AB) in the other and so on. Litters would be randomized to blocks as before, and the treatment combinations randomized to animals within litters. With this arrangement, quite a lot of information can be obtained concerning the interactions, because each interaction appears completely unconfounded in three replicates. Thus we have full information about ABC in replicates II, III and IV, and full information about AB in replicates I, III and IV and so on. Such interactions are then said to be *partially confounded*, and we indicate the extent of the confounding by the amount of information available, expressed as a fraction of the amount that would be obtained with this number of plots if there were no confounding, e.g. if there were no confounding in this experiment full information on each of the interactions would be obtained from 4 replicates, but in the confounded design it is obtained from only 3 replicates; so we say that $\frac{3}{4}$ of the relative information of ABC, AB, AC and BC is retained or $\frac{1}{4}$ lost. But from this it should not be assumed that these interactions are necessarily being tested less efficiently than they would be if the experiment was not a confounded one, because the very fact of dividing into small blocks reduces the error, in this case by removing most of the genetic variation from the error, so that very often the efficiency of the tests for interactions with only 3 replicates available would be greater than that for a test with 4 replicates not arranged in small blocks. The factors not confounded at all (A, B and C in this case) derive the full benefit of dividing into small blocks, since full information is obtained in all replicates for these, and they have relative information of 1.

The analysis of variance is shown in Table 9.3. We cannot easily break down the degrees of freedom for blocks as we did in the fully confounded case, but there is a degree of freedom available for each of the partially confounded interactions, because an unbiased comparison can be made in some of the replicates. We usually denote the partially confounded effects by primes, to remind ourselves that they need special care when finding standard errors and when building up tables for presenting the results.

The blocks sum of squares is obtained as usual, i.e. $\frac{1}{4}\sum L^2 - $ C.F., total S.S. is also got in the usual way, and A, B and C can be got in either of the ways shown previously for 2^n experiments when not confounded, e.g. by calculating the effects from the $+$ and $-$ table and calculating say $A^2/(r\sum k^2)$, in our case $A^2/(4 \times 8)$. The partially confounded effects are a little more tricky, but the general principal is that each must be calculated only from those blocks in which it is not confounded. This can be done directly, e.g. for AB' we need the totals over replicates I, III and IV for each treatment combination.

170 CONFOUNDING

			AB coefficients
(1)	$3·27 + 4·29 + 4·57$	= 12·13	+1
a	$3·96 + 4·22 + 4·83$	= 13·01	−1
b	$3·21 + 3·56 + 4·35$	= 11·12	−1
ab	$3·63 + 4·12 + 4·19$	= 11·94	+1
c	$3·64 + 4·36 + 4·66$	= 12·66	+1
ac	$3·67 + 3·80 + 3·92$	= 11·39	−1
bc	$3·15 + 4·25 + 3·90$	= 11·30	−1
abc	$4·06 + 4·34 + 4·89$	= 13·29	+1

Then applying the coefficients for AB we obtain $+3·20$ and the S.S. will be

$$\frac{3·20^2}{r\sum k^2} = \frac{3·20^2}{3 \times 8} = 0·4267$$

The divisor is only 3×8, not 4×8, since there are only 3 replications entering into this calculation.

We could do the same thing for AC' using only the replicates I, II and IV, for BC' using replicates I, II and III, and for ABC' using replicates II, III and IV. A little thought will indicate that there is a quicker method. Looking at the layout in Table 9.3 we see that the effect of AB in replicate II is the total of block II (ii) minus the total of block II (i). If we calculate the total effect of AB over all blocks first, we could obtain the total effect for AB in the blocks where it is not confounded (AB') by subtracting the difference between the totals for blocks II (ii) and II (i), i.e. using square brackets to denote totals or total effects.

$$[AB'] = [AB] - [\text{II(ii)}] + [\text{II(i)}] = +5·09 - 17·66 + 15·77 = +3·20 \text{ as before}$$
Likewise
$$[AC'] = [AC] - [\text{III(ii)}] + [\text{III(i)}] = -0·85 - 15·99 + 16·95 = +0·11$$
$$[BC'] = [BC] - [\text{IV(ii)}] + [\text{IV(i)}] = +2·45 - 18·19 + 17·12 = +1·38$$
$$[ABC'] = [ABC] - [\text{I(ii)}] + [\text{I(i)}] = +3·31 - 14·87 + 13·72 = +2·16$$

So the quickest way of doing the whole analysis is to get the block totals, treatment totals, and the effects over all blocks by the sums and differences method. Then correct the partially confounded effects for the blocks in which they are confounded. The full analysis of variance as in Table 9.3 can now be completed by calculating the S.S. of each unconfounded effect as its total effect squared and divided by 32, and that for each confounded effect as the corrected total effect squared and divided by 24. The error S.S. can be obtained as the difference between the total S.S. and the block and treatment sums of squares, and treatment effects tested by F-tests. If F-tests are not required, a simplified analysis of variance would be as shown:

	D.F.	S.S.	M.S.	
Blocks	7	$\dfrac{\Sigma L^2}{4} - \text{C.F.}$	$= 3\cdot 8982$	
Unconfounded effects	3	$\dfrac{[A]^2 + [B]^2 + [C]^2}{32}$	$= 0\cdot 6913$	
Partially confounded effects	4	$\dfrac{[AB']^2 + [AC']^2 + [BC']^2 + [ABC']^2}{24}$	$= 0\cdot 7009$	
Error	17	By difference	$= 1\cdot 2009$	$0\cdot 0706$
Total	31	$\Sigma x^2 - \text{C.F.}$	$= 6\cdot 4913$	

Standard errors for the total effects follow the usual rules: A, B and C are each the differences between totals of 16 observations, and so have an S.E. of $\sqrt{(2 \times 16 \times \text{M.S.}_E)} = \sqrt{2\cdot 2605} = 1\cdot 503$, whilst AB', AC', BC' and ABC' are differences between totals of 12 and so have an S.E. of $\sqrt{(2 \times 12 \times \text{M.S.}_E)} = \sqrt{1\cdot 6954} = 1\cdot 302$. When converting these total effects to mean effects, we must again remember the effective replication and divide the unconfounded by 16, and the partially confounded by 12. The S.E.s are divided likewise, i.e.

$$\frac{1\cdot 503}{16} = 0\cdot 0939 \quad \text{and} \quad \frac{1\cdot 302}{12} = 0\cdot 1085$$

The effects of treatments can then be tested by comparing each mean effect in the last column of the centre portion of Table 9.3 with its appropriate standard error by means of a t-test. If we wish to display the results in two-way tables, we usually cannot make up the tables from the raw data because, if we take means over the whole experiment, the values will be affected by block effects; if we exclude all blocks which affect any of the values in the table, we shall not be using all the information available. Thus all tables should be built up from the best estimates of the effects. For example, we know that in an unconfounded situation (p. 98) we can make an ab-table as

$$a_0 b_0 = M - \tfrac{1}{2}A - \tfrac{1}{2}B + \tfrac{1}{2}AB$$
$$a_1 b_0 = M + \tfrac{1}{2}A - \tfrac{1}{2}B - \tfrac{1}{2}AB$$
$$a_0 b_1 = M - \tfrac{1}{2}A + \tfrac{1}{2}B - \tfrac{1}{2}AB$$
$$a_1 b_1 = M + \tfrac{1}{2}A + \tfrac{1}{2}B + \tfrac{1}{2}AB$$

In the present case we can use our means over four replicates for estimates of A and B, and substitute our corrected mean AB' for AB in these formulae. This gives

$$a_0b_0 = 4{\cdot}071 - \tfrac{1}{2}(0{\cdot}283) - \tfrac{1}{2}(-0{\cdot}071) + \tfrac{1}{2}(0{\cdot}267) = 4{\cdot}10$$
$$a_1b_0 = 4{\cdot}071 + \tfrac{1}{2}(0{\cdot}283) - \tfrac{1}{2}(-0{\cdot}071) - \tfrac{1}{2}(0{\cdot}267) = 4{\cdot}11$$
$$a_0b_1 = 4{\cdot}071 - \tfrac{1}{2}(0{\cdot}283) + \tfrac{1}{2}(-0{\cdot}071) - \tfrac{1}{2}(0{\cdot}267) = 3{\cdot}76$$
$$a_1b_1 = 4{\cdot}071 + \tfrac{1}{2}(0{\cdot}283) + \tfrac{1}{2}(-0{\cdot}071) + \tfrac{1}{2}(0{\cdot}267) = 4{\cdot}31$$

Following the usual rule for variances (p. 77) the S.E. of each of these quantities will be

$$\sqrt{(\text{S.E.}_M^2 + \tfrac{1}{4}\text{S.E.}_A^2 + \tfrac{1}{4}\text{S.E.}_B^2 + \tfrac{1}{4}\text{S.E.}_{AB'}^2)}$$

which can be worked out directly or becomes

$$\sqrt{\left[\text{M.S.}_E\left(\frac{1}{32} + \frac{1}{4\times 8} + \frac{1}{4\times 8} + \frac{1}{4\times 6}\right)\right]} = \sqrt{\frac{13\,\text{M.S.}_E}{96}} = 0{\cdot}098$$

The whole table with marginal means and S.E.s would be:

	b_0	b_1	Mean	
a_0	4·10	3·76	3·93	
a_1	4·11	4·31	4·21	($\pm 0{\cdot}066$)
	($\pm 0{\cdot}098$)			
Mean	4·11	4·04		
	($\pm 0{\cdot}066$)			

It will be seen that when calculating the marginal means, AB' cancels out in both cases, so their S.E. is not

$$\frac{\text{S.E. for each cell}}{\sqrt{2}}$$

as it would be in the unconfounded case, but must be got separately as

$$\sqrt{(\text{S.E.}_M^2 + \tfrac{1}{4}\text{S.E.}_A^2)} = \sqrt{(\tfrac{1}{16}\text{M.S.}_E)}$$

Similarly, if we require to test some function of a confounded and a non-confounded effect, the ordinary rule of combining S.E.s applies, e.g. if we want confidence limits for the effects of a in the presence of b (i.e. $a_1b_1c_0 + a_1b_1c_0 - a_0b_1c_1 - a_0b_1c_0$ which without any confounding would be $A + AB$) we find the effect as $A + AB' = 0{\cdot}283 + 0{\cdot}267 = 0{\cdot}550$ and the S.E. must be

$$\sqrt{(\text{S.E.}_A^2 + \text{S.E.}_{AB'}^2)} = \sqrt{(0{\cdot}0939^2 + 0{\cdot}1085^2)} = \sqrt{0{\cdot}02059} = 0{\cdot}1435$$

This system can be extended to the confounding of several interactions when there are many more factors, but we need first to investigate another possibility of 2^n designs.

Single replicate experiments

Suppose we have five factors each at 2 levels, say A, B, C, D and E, making 2^5 or 32 combinations. We can divide up the degrees of freedom for treatments as:

	D.F.
Main effects (A, B, C, D, E)	5
Two-factor interactions ($AB, AC \ldots DE$)	10
Three-factor interactions ($ABC, ABD \ldots CDE$)	10
Four-factor interactions ($ABCD, ABCE \ldots BCDE$)	5
Five-factor interaction ($ABCDE$)	1

We can get the D.F. by actually specifying each interaction or more simply as

$$\frac{n!}{r!(n-r)!}$$

where n = total number of factors, and r = number of factors in each of the effects or interactions being considered.

Having divided up our treatment D.F. we see that only few of them are likely to represent effects of interest. In most cases three-factor and higher interactions are quite meaningless, and can be taken as nothing more than estimates of natural variation or estimates of the error that we have calculated in other experiments, and for this purpose they can be combined to give a complete analysis of variance of:

	D.F.
Main effects	5
Two-factor interactions	10
Error	16
Total	31

Each of the five main effects and each of the ten two-factor interactions would be worked out separately, and each tested against the error with 16 D.F. derived from the other interactions. Thus we have an experiment with no overall replication, which is a very useful saving when there is a large number of treatment combinations per replicate. The actual effects being tested have ample replication within this single replicate, e.g. each main effect is a comparison of 16 plots at the first level with 16 plots at the second level, whilst a two-factor interaction has eight replications of each combination in the 2×2 table from which it would be derived, both of which would be reckoned very good replication in, say, a randomized block design.

However, since single replicates can be used only when there is a large number of treatment combinations (32 is really the minimum) it is desirable to be able to arrange the experiment in blocks, and this can be done by confounding. From our arguments with the 2^3 we can easily see how to divide the 32 combinations into two blocks of 16 each. This would be done best by confounding the highest-order interaction $ABCDE$, though any interaction or indeed main effect could be chosen. We should then have an analysis of variance:

	D.F.
Blocks ($ABCDE$)	1
Main effects	5
Two-factor interactions	10
Error	15
Total	31

But 16 plots per block is still too large a number for many experiments, so we would like to divide them further. Suppose we tried to get 8 combinations per block, there would then be 4 blocks, and so 3 D.F. of interactions must be confounded. We could first divide into groups of 16 according to one chosen interaction, and then divide each of the groups into two blocks of 8, according to another interaction. Suppose we chose AB for the first split; we should have (1) and ab in all combinations with c, d and e in one group and a and b in all combinations with c, d and e in the second group. Then we might divide each group into two blocks according to the CD-interaction, when within the first group the first block would contain (1), ab, cd and $abcd$ in all combinations with e, whilst the second would contain c, abc, d and abd in all combinations with e. Likewise in the second group the same c and d-combinations would appear with a and b. Another way of looking at it is to use the plus and minus table (we can omit e since each of these will appear with and without e):

	(1)	a	b	ab	c	ac	bc	abc	d	ad	bd	abd	cd	acd	bcd	$abcd$
AB	+	−	−	+	+	−	−	+	+	−	−	+	+	−	−	+
CD	+	+	+	+	−	−	−	−	−	−	−	−	+	+	+	+

Now + + in a column forms the first block, + − the second, − + the third and − − the fourth or, in full, we have:

Blocks	I (+ +)	II (+ −)	III (− +)	IV (− −)
	(1)	c	a	ac
	ab	abc	b	bc
	cd	d	acd	ad
	$abcd$	abd	bcd	bd
	e	ce	ae	ace
	abe	$abce$	be	bce
	cde	de	$acde$	ade
	$abcde$	$abde$	$bcde$	bde

We have four blocks and they must have 3 D.F. so, in confounding AB and CD, we must have confounded another effect. To find out which, let us consider the breakdown of the block D.F.

Blocks	I	II	III	IV
AB	+1	+1	−1	−1
CD	+1	−1	+1	−1

We remember that if four items are divided into two contrasts in this way, the contrast which is orthogonal to these two (so that the products of every pair of sets of coefficients sum to zero) is, in the order above, $+1-1-1+1$ which is the product of the coefficients for AB and CD, and consequently is the interaction between AB and CD or $ABCD$. To check, look at the completed layout; all combinations in both Blocks I and IV contain an even number of the letters a, b, c and d, and those in Blocks II and III an odd number of those letters. This is a perfectly general finding in 2^n experiments: if we divide into two groups, we confound one effect to do so; if we divide into four groups, we do so by confounding two effects, but also confound their interaction in the process; if we divide into eight groups, we must use three effects but confound as well all their interactions in the process, e.g. confounding AB, CD, EF involves also confounding $ABCD$, $ABEF$, $CDEF$ and $ABCDEF$, seven in all, as we might expect if there are eight blocks. This seems obvious enough when the interactions confounded are distinct in this way, but what would happen if we had chosen AB and BC for our confounding? What is the interaction $AB \times BC$? Going back to the table coefficients we have:

	(1)	a	b	ab	c	ac	bc	abc	...all combinations with d and e
AB	+	−	−	+	+	−	−	+	
BC	+	+	−	−	−	−	+	+	
$AB\times BC$	+	−	+	−	−	+	−	+	

which we see is the same as AC (the $+$s contain an even number of the letters a, c). This is obvious if we remember that $AB \times BC = A \times B \times B \times C$ and $B \times B$ will be either $(+1)\times(+1)$ or $(-1)\times(-1)$, i.e. $+1$ in every case; so $A \times B \times B \times C = A \times 1 \times C = A \times C$. So here is another general rule for 2^n experiments; the interaction between two interactions is the combination of all the letters appearing in the interactions with any letters which appear twice struck out, i.e. $AB \times BC = ABBC = AC$; likewise $ABCD \times BCDE = ABCDBCDE = AE$. Such interactions are usually referred to as *generalized interactions*.

Returning to our single replication of 2^5 plots, we would need to be careful in choosing which interactions to confound. If we chose the highest order *ABCDE* as one, and one of the next highest order as the other, say *ABCD*, then the generalized interaction of these is *E* and we should have confounded a main effect, which is usually not very helpful. We can only use the five-factor interaction when dividing into blocks of eight combinations if we are prepared to lose information on one two-factor interaction; then we can have *ABCDE* and *ABC* with generalized interaction *DE*. There are 10 such sets, since any two-factor interaction may be chosen, but the most useful type of set to use is a four-factor interaction, with a three-factor one containing the letter omitted in the four-factor. We then have *BCDE*, *ABC*, *ADE* and there are 15 such sets, since there are 5 four-factor interactions, each of which can be divided into pairs in 3 ways.

The analysis of variance is

	D.F.
Blocks	3
Main effects	5
Two-factor interactions	10
Error	13

These principles can be applied to any 2^n experiment, where n is greater than 5, to get single replicate experiments with blocks of reasonable size, and many useful plans are given in Yates (1937) and in Cochran and Cox (1957).

As an example of the procedure, suppose a nutritionist wishes to study in one experiment the effects of the following treatments on live-weight gain and efficiency of food conversion from weaning to bacon weight in pigs:

- (*A*) two feeding scales;
- (*B*) two protein levels in the diet;
- (*C*) feeding wet or dry food;
- (*D*) an antibiotic supplement at the two levels of 0 and 7 kg/t meal;
- (*E*) a mineral supplement at the two levels of 0 and 0·5 per cent.

He assumes that sex of pig may be important, and that he cannot conveniently obtain from any one litter more than four pigs of each sex.

There are six factors, each at two levels, if we include sex (call it *F*). If we can expect four of each sex in each litter, we can have groups of 8 provided *F* is not confounded. Therefore we might have 2^6 treatments in blocks of 8, requiring $2^6 \div 8 = 8$ blocks per replicate. So 7 D.F. must be confounded with blocks.

CONFOUNDING

The skeleton analysis of variance of the treatments is

	D.F.
Main effects	6
Two-factor interactions	15
Three-factor interactions	20
Four-factor interactions	15
Five-factor interactions	6
Six-factor interaction	1
Total	63

We require to confound three "primary" interactions and their four generalized interactions. We cannot use the six-factor interaction with a five- or four-factor one without confounding a main effect or two-factor interaction. Using the six-factor interaction with a three-factor one gives a three-factor generalized interaction, e.g.

$$ABCDEF \text{ and } ABC \text{ gives } DEF$$

but any other three-factor must contain two letters of either ABC or DEF and so would lead to a two-factor interaction being confounded.

Confounding two five-factor interactions leads to a two-factor interaction being confounded. Confounding a five-factor and a four-factor interaction can lead to a three-factor interaction being confounded, e.g.

$$ABCDE \text{ and } ABCF \text{ gives } DEF \text{ as generalized interaction}$$

Any additional four-factor must contain F if a main effect is not to be confounded, and must contain only one of the letters ABC if its interaction with $ABCF$ is not to lead to a two-factor interaction being confounded. Therefore it must contain all the letters DEF and so confound a main effect when interacting with DEF.

Using a four-factor interaction, say $ABCD$, we cannot use another four-factor one containing three of the same letters, but using one with two of the same letters, we get, e.g., $ABCD$ and $ABEF$ with $CDEF$ as generalized interaction.

We cannot use another four-factor following this rule, say $ACEF$, since it will give a two-factor generalized interaction with one of the other interactions. However, combining these three with a three-factor interaction is possible. The three-factor must contain two letters of each of the four-factor interactions; if it has three, it would confound a main effect when interacting with one of them. Therefore *ACE* is possible giving

ABCD
ABEF
CDEF
ACE
BDE
BCF
ADF

To determine which combinations go in each block (or on each litter in this case) we use the system of writing the factor letter for the higher level of each factor, and no letter for the lower level, and choosing three of the lowest-order interactions to be confounded, such that no one is the generalized interaction of the other two, first find the block which contains the combinations which have an even number of letters of all three (this is the block which contains (1)). In this case taking *ACE*, *BDE* and *BCF*, since all the letters appear in the interactions, no single letter combination can be in this block, nor can any pairs of letters since, if a pair appeared in two three-factor interactions, the interaction between these two would be a two-factor one. This we have avoided, and if we had a pair in one interaction, neither of which appeared in the other two, the latter must themselves contain a common pair. Possible three-letter combinations can be obtained easily by inspection, thus *acf* has either two or no letters in the three interactions and is in the block containing (1). Likewise *ade*, *bce* and *bdf*. Similarly with four-letter combinations *abcd*, *abef* and *cdef*, which makes the total of eight combinations that we sought. These are written as the first column of a table and then, multiplying each by a combination which has not appeared, and omitting any letters which appear twice, will give another block. This process is continued until all eight blocks have been obtained, e.g.

Column 1	$1 \times a$	$1 \times b$	$1 \times c$	$1 \times d$	$1 \times e$	$1 \times f$	$1 \times ab$
(1)	a	b	c	d	e	f	ab
acf	cf	abcf	af	acdf	acef	ac	bcf
ade	de	abde	acde	ae	ad	adef	bde
bce	abce	ce	be	bcde	bc	bcef	ace
bdf	abdf	df	bcdf	bf	bdef	bd	adf
abcd	bcd	acd	abd	abc	abcde	abcdf	cd
abef	bef	aef	abcef	abdef	abf	abe	ef
cdef	acdef	bcdef	def	cef	cdf	cde	abcdef

Fractional replication

When we need experiments with six or more factors, although we can by confounding arrange for small blocks, there is another thing to consider. We probably want to measure the main effects and the two factor interactions; that would be the reason for choosing a factorial arrangement, but we may not need such a high degree of replication of these effects as the single replicate will give us. For instance, in a 2^6 design there will be 32-fold replication of main effects. Also the experiment provides information about higher-order interactions which is unnecessary, and even when most of these interactions are used to assess the error, such error may be obtained much more precisely than the treatments merit. So we might think that we could use our time much more efficiently if we had a smaller experiment. This is possible using what is known as *fractional replication*. In previous examples we have lost information about interactions by confounding them with blocks, now we will leave them out altogether. To see what is involved, let us look again at the 2^3 design, though we should never have a fractional replicate of this. Suppose we were prepared to lose the ABC-interaction altogether, then we could include in the experiment only the combinations that have the positive or the negative coefficients when calculating that interaction, i.e. either (1), ab, ac and bc, or a, b, c and abc. Let us take the second and then write coefficients for the various effects:

	a	b	c	abc	
A	$+1$	-1	-1	$+1$	
B	-1	$+1$	-1	$+1$	
C	-1	-1	$+1$	$+1$	
AB	-1	-1	$+1$	$+1$	$= C$
AC	-1	$+1$	-1	$+1$	$= B$
BC	$+1$	-1	-1	$+1$	$= A$
ABC	$+1$	$+1$	$+1$	$+1$	$=$ Total

So in an analysis we cannot distinguish between C and AB, B and AC, A and BC, or ABC and the total. C and AB, B and AC, etc., are said to be *aliases*. Thus, to use this design we must be able to arrange that the factors or interactions that we are interested in all have as aliases only interactions which we can reasonably assume have no true effect. Then any effect we measure can be said to be due to the factor we are interested in. To design such an experiment we first choose an interaction to be lost altogether; for a half replicate this will usually be the highest-order interaction, e.g. $ABCDEF$ in 2^6, and this is called the *defining contrast*. Now we can write down the aliases, which are simply the generalized interaction between the effects and the defining contrast thus:

$A = BCDEF$	$AB = CDEF$	$ABC = DEF$
$B = ACDEF$	$AC = BDEF$	$ABD = CEF$
$C = ABDEF$	$AD = BCEF$	$ABE = CDF$
$D = ABCEF$	$AE = BCDF$	$ABF = CDE$
$E = ABCDF$	$AF = BCDE$	$ACD = BEF$
$F = ABCDE$	$BC = ADEF$	$ACE = BDF$
	$BD = ACEF$	$ACF = BDE$
	$BE = ACDF$	$ADE = BCF$
	$BF = ACDE$	$ADF = BCE$
	$CD = ABEF$	$AEF = BCD$
	$CE = ABDF$	
	$CF = ABDE$	
	$DE = ABCF$	
	$DF = ABCE$	
	$EF = ABCD$	

In this case we would be well satisfied, as all main effects have five-factor interactions as aliases, and all two-factor interactions have four-factor ones, none of which could we ever conceive of having true effects. We decide by random numbers whether to use the combinations with positive or those with negative coefficients for the defining contrast, and then write down the actual treatment combinations to be used by one of the methods we have used before—in fact the simplest is to write odd or even numbers of the letters contained in the defining contrast. Thus, if the combinations with positive coefficients were chosen, they must include *abcdef*, an even number of letters, so we can write out all the treatments as:

(1), *ab*, *ac*, *ad*, *ae*, *af*, *bc*,...*ef*, *abcd*, *abce*,... *cdef*, *abcdef*

Or, if the combinations with negative coefficients were chosen, *abcdef* must not appear, so we write the odd combinations:

a, *b*, *c*, *d*, *e*, *f*, *abc*, *abd*,... *def*, *abcde*, *abcdf*,... *bcdef*

Treatment combinations are randomized to plots in the usual way.
The analysis of variance would be:

	D.F.
Main effects	6
Two-factor interactions	15
Error	10
Total	31

Applying the formula

$$\frac{n!}{r!(n-r)!}$$

it is found that a 2^6 design has 6 main effects, 15 two-factor interactions, 20 three-factor interactions, 15 four-factor interactions, 6 five-factor interactions, and 1 six-factor interaction. The six-factor interaction does not appear in the analysis because it is the defining contrast; the main effects have the 6 five-factor interactions as aliases, and the 15 two-factor interactions have the 15 four-factor ones as aliases, so the residual or error is simply the 20 three-factor interactions taken in pairs, i.e. $ABC = DEF$, $ABD = CEF$, etc., 10 pairs in all. Although we can get the sum of squares for error by difference in the usual way, it is often simpler to obtain it directly by squaring the appropriate effects, summing the squares, and dividing by the appropriate divisor. All effects can be calculated using the usual table of coefficients or by a modification of the method of sums and differences.

In our example we have produced a half-replicate, but with a larger number of factors (e.g. 8) it is possible to get a quarter replicate with all main effects and two-factor interactions having only high-order interactions as aliases. With larger numbers still, we can get $\frac{1}{8}$, $\frac{1}{16}$ and so on replication. The procedure is exactly the same as in confounding. To obtain $\frac{1}{4}$ replicate we must choose two defining contrasts, and their generalized interaction will also be a defining contrast, e.g. with 8 factors we might choose $ABCDE$ as one and $DEFGH$ as the other, then $ABCFGH$ would also be a defining contrast. Consequently each effect now has three aliases, e.g. $A = BCDE = ADEFGH = BCFGH$, so great care is needed in choosing the defining contrasts, e.g. if we chose $ABCDEFGH$ and $ABCDEF$, the remaining defining contrast would be GH and so H would be one of the aliases of G.

We can also arrange fractional replicates in blocks by confounding any interaction other than the defining contrasts. But we must remember that not only will the chosen interaction be confounded but so will its alias. Thus in our half replicate of 2^6 we could arrange it in two blocks of 16, by confounding a three-factor interaction, say ABC, and its alias DEF would be confounded, whereas if we chose $ABCD$, then its alias EF, which is one of the things we wanted to test, would be confounded and lost.

Fractional replication of 2^n designs is a very useful device for biologists in the early stages of investigations, when there is a large number of possible factors involved, which can interact with each other. The first requirement is usually to screen this array of factors to see which are worthy of future study. In this situation two levels, presence and absence, usually suffice and only large differences matter, work continuing on a factor if it has a sizeable effect on average, or if it modifies the effect of another factor. An example of such use is in investigation of deficiency diseases in plants, where there are many elements which are possible candidates for causing the symptoms. Cochran and Cox (1957, Chapter 6A) gives lists of useful plans for fractional replication with various sizes of blocks.

Confounding with three-level factors

As aids to design we have confounding, partial confounding, single replicates, and fractional replication, all of which are quite easily obtained when there are 2^n combinations. The principles also apply to factors at more than two levels, say p^n, but it is often more laborious to sort out the interactions. As an example of the sort of approach, we will consider 3^n. Taking the simplest member, 3^2 which has 9 treatment combinations, it is clear that we cannot confound an interaction with a single degree of freedom, because there is an odd number of treatments, and we cannot have $4\frac{1}{2}$ plots per block. In fact the only possible balanced division is into three blocks of three plots each, and this requires 2 D.F. for blocks, so we must find an interaction with 2 D.F. The clue to how this may be done has come to different people in different ways, but the one some appreciate is this: when we split up a 2×2 arrangement of treatments we considered the 4 combinations to form a square.

	b_0	b_1
a_0	00	01
a_1	10	11

(In this notation the first digit in each cell refers to the level of the factor forming the rows, and the second to the level of the factor forming the columns.) The 3 D.F. were split into one for rows which represented a, one for columns representing b, and one for diagonals $(00+11-01-10)$ representing interaction. A 2×2 square has two diagonals, hence 1 D.F.

A 3×3 square has two sets of three diagonals which we might call I and J.

	b_0	b_1	b_2
a_0	00	01	02
a_1	10	11	12
a_2	20	21	22

$$\begin{aligned} I_1 &= 00+11+22 \\ I_2 &= 10+21+02 \\ I_3 &= 20+01+12 \end{aligned} \quad \text{and} \quad \begin{aligned} J_1 &= 00+21+12 \\ J_2 &= 10+01+22 \\ J_3 &= 20+11+02 \end{aligned}$$

Now we note that each I contains one element only from each row, and one only from each column, so the Is are orthogonal to both rows and columns. Likewise the Js are orthogonal to rows and columns. Furthermore each I contains one element and only one from each J, and vice versa, so the Is and Js are mutually orthogonal. So now we could set out the full analysis:

	D.F.
Rows = A	2
Columns = B	2
Interaction $AB = \begin{cases} I \\ J \end{cases}$	2 2
Total	8

If we wanted to divide the nine treatment combinations into blocks of three, we could confound either the I or the J of the interaction with blocks and so have, e.g.

	D.F.
Blocks (I)	2
A	2
B	2
J	2

the blocks being

Block 1:	a_0b_0	a_1b_1	a_2b_2
Block 2:	a_1b_0	a_2b_1	a_0b_2
Block 3:	a_2b_0	a_0b_1	a_1b_2

When divided in this way with the I confounded in every replicate, we lose all practical information about the interaction AB, since J by itself or I by itself has no real meaning in terms of AB. However, if there are a number of replicates, we can confound I in half of them and J in the other half, and have half the information on AB by calculating the Is and Js from the replicates in which each is not confounded; these can be tested in an analysis of variance, and a table of yields of individual combinations free of block effects can be constructed, e.g.

$$a_1 b_2 = \text{mean} + \text{dev } A_1 + \text{dev } B_2 + \text{dev } I'_3 + \text{dev } J'_1$$

where dev A_1 means the difference between the mean of all the a_1s and the general mean, and dev I'_3 the difference between the mean of I_3 and the general mean in the replicates in which I is not confounded, and so on. (When there are more than two levels, dev A_1, standing for "deviation from the mean due to A_1, replaces the $\frac{1}{2}A$ in the formulae for 2^n designs; in that case $-\frac{1}{2}A = \text{dev } A_0$ and $+\frac{1}{2}A = \text{dev } A_1$, since A is the difference between the means of a_1 and a_0, and half this difference is the difference between either one of them and the mean of the two.)

This confounding of 3^2 is not used very often, but 3^3 experiments are very commonly required. In agriculture, much of the basic work on manuring of crops is done using the three fertilizer nutrients, N, P and K, and three levels are required to get some idea of the optimum dressing, so we might look at this in more detail. The common arrangement is a single replicate arranged in three blocks of nine plots. (It is also useful for experiments with pigs where we can have three litters of nine pigs each.) Now the 3^3 or 27 combinations give the effects:

	D.F.
A	2
B	2
C	2
AB	4
AC	4
BC	4
ABC	8

and usually we want full information concerning A, B and C, and quite a lot of information concerning AB, AC and BC, but ABC is of little value. Hence we can confound a part of ABC with blocks, and use the rest for error. Now if we remember that AB could be sub-divided to give

	D.F.
I_{AB}	2
J_{AB}	2

ABC, which is $C \times AB$, can be sub-divided as

	D.F.
$C \times I_{AB}$	4
$C \times J_{AB}$	4

and each of these forms a 3×3 table, e.g.

$$
\text{of } AB \left\{ \begin{array}{c|ccc} & c_0 & c_1 & c_2 \\ I_1 & 10 & 11 & 12 \\ I_2 & 20 & 21 & 22 \\ I_3 & 30 & 31 & 32 \end{array} \right.
$$

and can be sub-divided further to give

	D.F.
I of $C \times I_{AB}$	2
J of $C \times I_{AB}$	2
I of $C \times J_{AB}$	2
J of $C \times J_{AB}$	2

and one of these can be chosen at random to be confounded with blocks. Suppose we choose the first. The three groups required are $(I_1 c_0 + I_2 c_1 + I_3 c_2), (I_2 c_0 + I_3 c_1 + I_1 c_2)$ and $(I_3 c_0 + I_1 c_1 + I_2 c_2)$ which, putting the levels of a and b appropriate to each I, is easily translated back to individual combinations of abc and becomes

Block 1 $000 + 110 + 220 + 101 + 211 + 021 + 202 + 012 + 122$
Block 2 $100 + 210 + 020 + 201 + 011 + 121 + 002 + 112 + 222$
Block 3 $200 + 010 + 120 + 001 + 111 + 221 + 102 + 212 + 022$

to give us the sets of combinations which should be randomized to the plots in the appropriate blocks after sets have been randomized to blocks.

In fact there is a simpler way of finding the sets if we are not concerned to know which interaction we have confounded; that is to write down any 3×3 Latin square chosen at random, using numbers instead of letters:

		b		
		0	1	2
	0	0	1	2
a	1	1	2	0
	2	2	0	1

Let these be the levels of c; put levels of a to represent rows and of b to represent columns, and we now have the nine combinations to represent one block of the experiment, i.e. $000\quad 011\quad 022\ldots 221$; then make a cyclical change of numbers in the Latin square, $0 \to 1$, $1 \to 2$, $2 \to 0$ and write down the second block; the remainder is the third block.

The analysis of variance becomes

Source of variation	D.F.
Blocks	2
Main effects	6
Two-factor interactions	12
Error	6
Total	26

which gives low precision for the error term if only one replicate is used. Frequently it is desirable to carry out the experiment on a number of sites in the case of a fertilizer investigation, or on a number of occasions in other biological investigations; then, provided the variance is the same from site to site, or occasion to occasion, the error terms can be combined to give more precise estimates.

When using a single replicate for exploratory purposes, only a few null hypotheses may be involved, and it may be reasonable to assume that parts of two-factor interactions represent only natural variation. For example, we might be concerned to know if there were linear effects, and deviations from linear effects due to each of the three factors, and if there were linear × linear interactions between each pair of factors. Then, assuming that linear × quadratic and quadratic × quadratic interactions estimate nothing but error variation, we could have an analysis of variance of

Source of variation	D.F.
Blocks	2
A linear (A')	1
A quadratic (A'')	1
B'	1
B''	1
C'	1
C''	1
$A' \times B'$	1
$A' \times C'$	1
$B' \times C'$	1
Error	15
Total	26

which gives good precision for the error term. Sometimes it might be desirable to take out linear × quadratic interactions, leaving 9 D.F. for error, but it must be stressed that the analysis to be adopted must be decided in the light of the hypotheses made before the experiment starts. It is pure deception to do the analysis in all possible ways and choose the one that suits the experimenter's ideas the best.

Confounding with factors at more than three levels

Looking back at our I and J divisions, we can see similarity to Latin squares. If we write $I_1 = A$, $I_2 = B$, $I_3 = C$; $J_1 = \alpha$, $J_2 = \beta$ and $J_3 = \gamma$, we can combine the two squares into

	a_0	a_1	a_2
b_0	$A\alpha$	$C\beta$	$B\gamma$
b_1	$B\beta$	$A\gamma$	$C\alpha$
b_2	$C\gamma$	$B\alpha$	$A\beta$

and we find as expected that each Latin letter appears once, and once only, in each row and column, and with each Greek letter. Likewise each Greek letter appears once, and once only, in each row and column and with each Latin letter. We have what is known as a Graeco-Latin square, one of a family known as *orthogonal Latin squares*. It should now give us a clue as to how to proceed with confounding when factors have more than three levels. Suppose we have 4×4 treatments. We could write

	0	1	2	3
0	$Ad\gamma$	$Bb\alpha$	$Ca\delta$	$Dc\beta$
1	$Ba\beta$	$Ac\delta$	$Dd\alpha$	$Cb\gamma$
2	$Cc\alpha$	$Da\gamma$	$Ab\beta$	$Bd\delta$
3	$Db\delta$	$Cd\beta$	$Bc\gamma$	$Aa\alpha$

a hyper-Graeco-Latin square in three alphabets. We could make blocks of four combinations using the capitals, using the lower-case Latin letters, or using the Greek letters. They would be mutually orthogonal and orthogonal to the main effects. With 3 replicates a balanced design using one set for each replicate enables $\frac{2}{3}$ information on the interaction to be obtained. Similar arrangements can be had for any number of levels for which the necessary orthogonal Latin squares exist. This is for most useful numbers except 6.

The Graeco-Latin square is a classical design in itself, for use when there are two sets of treatments and no interaction between them. The analysis of variance is

Source of variation	D.F.
Rows	$r-1$
Columns	$r-1$
Latin letters	$r-1$
Greek letters	$r-1$
Error	$(r-1)(r-3)$
Total	r^2-1

These designs are often useful in experiments with tree crops which take a long time to establish, so some early treatments are first applied and then another set of treatments when the trees are fully grown, thus avoiding the need to establish trees specially for the second experiment.

Confounding when all factors do not have the same number of levels

It remains now to consider if there are any possibilities of reducing block size when all the factors do not have the same number of levels. Then it is usually impossible to divide the combinations into sets which correspond to a single high-order interaction, and we have to be satisfied with partially confounding a number of interactions; these may not be completely free of confounding in any particular replicate, so treatment effects have to be corrected for block effects. The best designs are those which confound the more important interactions the least. We will look at an experiment in which factor A is at three levels a_0, a_1, a_2, and factors B and C each at two levels, b_0 and b_1, c_0 and c_1, giving $3 \times 2 \times 2 = 12$ combinations in all.

To keep the main effects of A unconfounded, we must have each a represented the same number of times in each block, so blocks must contain a multiple of 3 plots. Likewise to keep B and C unconfounded there must be the same number of b_0 as b_1 and of c_0 as c_1 in each block, so blocks must contain a multiple of 2 plots. Hence to keep A, B and C unconfounded, there must be 6 or a multiple of 6 plots per block and, since we have only 12 combinations, blocks of 6 are the only ones possible. Now if we have 6 plots per block, we can arrange for each of the six combinations of ab and each of the six combinations of ac to be present in each block; so A, B, C, AB and AC can all be left unconfounded. BC, however, comes from 4 combinations, and we cannot arrange blocks of six so that they contain each of these combinations the same number of times; so BC must be partially confounded, and of course ABC involving 12 combinations must also be confounded. To see how best to arrange the blocks, let us write the complete set of combinations for ab and then put in the cs in various ways which give all combinations of ac.

	I		II		III		IV	
ab	(i)	(ii)	(i)	(ii)	(i)	(ii)	(i)	(ii)
				c				
00	1	0	0	1	0	1	0	1
01	0	1	1	0	1	0	1	0
10	0	1	1	0	0	1	0	1
11	1	0	0	1	1	0	1	0
20	0	1	0	1	1	0	0	1
21	1	0	1	0	0	1	1	0

CONFOUNDING

If we used set IV, we should find that the BC interaction was completely confounded with blocks, i.e. all the b_0c_0 and b_1c_1 are in IV(i) and all the b_0c_1 and b_1c_0 are in IV(ii); but in the other three sets it appears once in each block (half of it appears a second time, so it lacks orthogonality). An example of using these three sets is shown in Table 9.4.

Table 9.4—A $3 \times 2 \times 2$ design with partial confounding

Yields

Blocks	I(i)	I(ii)	II(i)	II(ii)	III(i)	III(ii)	Total
Treatments							
a b c							
0 0 0		1·59	1·87		1·80		5·26
0 0 1	1·89			2·40		1·93	6·22
0 1 0	1·59			2·73		1·76	6·08
0 1 1		1·75	1·89		2·13		5·77
1 0 0	1·71			2·41	1·67		5·79
1 0 1		1·76	1·69			1·82	5·27
1 1 0		1·56	1·88			1·92	5·36
1 1 1	1·81			3·01	1·88		6·70
2 0 0	1·69		1·91			1·72	5·32
2 0 1		1·62		2·57	1·85		6·04
2 1 0		1·63		2·89	1·79		6·31
2 1 1	1·80		1·88			2·06	5·74
Total	10·49	9·91	11·12	16·01	11·12	11·21	69·86
g	−0·58		+4·89		+0·09		

Total effects over all replicates

	A_0	A_1	A_2	Total
Total	23·33	23·12	23·41	69·86
B	+0·37	+1·00	+0·69	+2·06
C	+0·65	+0·82	+0·15	+1·62
BC	−1·27	+1·86	−1·29	−0·70

C.F. = 135·5672

Analysis of variance

Source of variation	D.F.	S.S.	M.S.
Blocks	5	4·0230	
A	2	0·0038	0·0019
B	1	0·1179	0·1179
AB	2	0·0165	0·0083
C	1	0·0729	0·0729
AC	2	0·0202	0·0101
BC'	1	0·0184	0·0183
ABC'	2	0·0244	0·0122
Error	19	0·3655	0·0192
Total	35	4·6626	

The sums of squares for blocks, total and all unconfounded effects are obtained in the usual way. To get the BC' sum of squares it is best first to get the ordinary interaction total in the usual way, i.e. $b_0c_0 + b_1c_1 - b_0c_1 - b_1c_0$ over all levels of a, and then correct it for the block effects. It will be seen that in every replicate there are two extra plots of the positive side of the interaction in every block labelled (i), and two extra of the negative side in every block labelled (ii). Therefore any difference between the blocks (i) and (ii) of any pair will add to the apparent effect of BC, so such differences must be subtracted, which is the same as adding the difference of block (ii) and block (i). Thus to correct for this we work out the differences between the block totals for each replicate taking

$$g_1 = \text{I(ii)} - \text{I(i)}$$
$$g_2 = \text{II(ii)} - \text{II(i)}$$
$$g_3 = \text{III(ii)} - \text{III(i)}$$

Since there are 6 plots in a block, each plot will be affected by $\frac{1}{6}g$, and since we have two plots per block affected, our best estimate of the total BC' effect would be

$$BC + \tfrac{1}{3}(g_1 + g_2 + g_3)$$

where BC is the total effect calculated over all replicates. To avoid the fractions we usually work out

$$3Q = 3BC + g_1 + g_2 + g_3$$

and the S.S. for $3Q$ would be

$$\frac{(3Q)^2}{\sum k^2}$$

where the k are the coefficients applied to individual observations to obtain $3Q$. In earlier examples, each individual value contributed in only one way to the quantity which was squared to obtain an S.S., so the k of each was obvious. Now we see that each value will contribute to two parts of the expression for deriving $3Q$, i.e. to BC and to one of the gs. Its total contribution (i.e. its combined coefficient) can be obtained only by considering the whole function. We can set this out and put in the coefficients for each part for each value, and sum them to get the final k. Doing this for the plots of Block I(i) in order gives

$3BC$	$+g_1$	$+g_2$	$+g_3$	=	$3Q$
-3	-1	0	0	=	-4
-3	-1	0	0	=	-4
$+3$	-1	0	0	=	$+2$
$+3$	-1	0	0	=	$+2$
$+3$	-1	0	0	=	$+2$
$+3$	-1	0	0	=	$+2$

All blocks are similar so $\sum k^2 = 6(2 \times 4^2 + 4 \times 2^2) = 288$.

If unconfounded $3BC$ would have $\sum k^2 = 36 \times 3^2 = 324$ so we have $\frac{288}{324}$ of the information or have lost $\frac{36}{324} = \frac{1}{9}$. The remainder of the loss must be in ABC. If we had no confounding we could get the BC-interaction when A had the level a_0 and when it had a_1 and a_2. Then the S.S. for ABC would be

$$\frac{BC_{a_0}^2 + BC_{a_1}^2 + BC_{a_2}^2}{\sum k^2} - \frac{BC^2}{3 \sum k^2}$$

but now we correct each of these BC-terms for blocks, subtracting or adding the difference between blocks according to which combinations of bc they contain, in the same way as before.

Again to avoid fractions we get

$$3R_0 = 3BC_{a_0} - g_1 + g_2 + g_3$$
$$3R_1 = 3BC_{a_1} + g_1 - g_2 + g_3$$
$$3R_2 = 3BC_{a_2} + g_1 + g_2 - g_3$$

To get the divisor we have to extend our previous rule. When we have to square a single quantity to get an S.S., we divide by the sum of squares of the coefficients; if we have to square several quantities, we again divide by the sum of squares of the coefficients of each, but *only* provided no individual value appears in more than one of them. If the same value or same plot appears in more than one, as it does here, where almost all plots contribute to each R, then to find the divisor we must find the coefficients for the difference between two of the quantities, say $3R_0 - 3R_1$, and the divisor is half the sum of the squares of these coefficients.

Taking the plots in the order given in Table 9.4 we have

Block I (i)

$3BC_{a_0} - g_1 + g_2 + g_3$					$-3BC_{a_1} - g_1 + g_2 - g_3 =$			$3R_0 - 3R_1$
-3	$+1$	0	0	0	$+1$	0	0	$= -1$
-3	$+1$	0	0	0	$+1$	0	0	$= -1$
0	$+1$	0	0	-3	$+1$	0	0	$= -1$
0	$+1$	0	0	-3	$+1$	0	0	$= -1$
0	$+1$	0	0	0	$+1$	0	0	$= +2$
0	$+1$	0	0	0	$+1$	0	0	$= +2$

Block I (ii) will be likewise with the signs changed. Blocks II (i) and II (ii) will also have the same coefficients. Block III (i) will have

$$
\begin{array}{ccccccccc}
& 3BC_{a_0} & -g_1 & +g_2 & +g_3 & -3BC_{a_1} & -g_1 & +g_2 & -g_3 = & 3R_0 - 3R_1 \\
+3 & & 0 & 0 & -1 & 0 & & 0 & 0 & +1 & = +3 \\
+3 & & 0 & 0 & -1 & 0 & & 0 & 0 & +1 & = +3 \\
0 & & 0 & 0 & -1 & -3 & & 0 & 0 & +1 & = -3 \\
0 & & 0 & 0 & -1 & -3 & & 0 & 0 & +1 & = -3 \\
0 & & 0 & 0 & -1 & 0 & & 0 & 0 & +1 & = 0 \\
0 & & 0 & 0 & -1 & 0 & & 0 & 0 & +1 & = 0 \\
\end{array}
$$

Block III (ii) will be the same with the signs changed.

$$
\begin{aligned}
\text{Thus } \sum k^2 \text{ is } 2 \times 4 \times 9 &= 72 \\
4 \times 2 \times 4 &= 32 \\
4 \times 4 \times 1 &= 16 \\
\hline
120 &
\end{aligned}
$$
and the divisor we seek is 60

Without confounding $\sum k^2 = 6 \times 2 \times 9 = 108$. So $\frac{60}{108}$ of information is retained or $\frac{48}{108} = \frac{4}{9}$ lost on each degree of freedom of the ABC-interaction. There are two degrees of freedom, so total loss for $ABC = \frac{4}{9} + \frac{4}{9} = \frac{8}{9}$ and this together with the $\frac{1}{9}$ loss for BC makes unity.

For the example in Table 9.4 to obtain the S.S. for BC' we need

$$
\begin{aligned}
3Q &= 3BC + g_1 + g_2 + g_3 \\
&= 3 \times (-0.70) - 0.58 + 4.89 + 0.09 = +2.30
\end{aligned}
$$

and the S.S. is

$$
\frac{(+2.30)^2}{288} = 0.0184
$$

For ABC' we need

$$
\begin{aligned}
3R_0 &= 3(-1.27) + 0.58 + 4.89 + 0.09 = +1.75 \\
3R_1 &= 3(+1.86) - 0.58 - 4.89 + 0.09 = +0.20 \\
3R_2 &= 3(-1.29) - 0.58 + 4.89 - 0.09 = +0.35
\end{aligned}
$$

Note that $\sum(3R) = 3Q$, which is a useful check on the arithmetic and

$$
\text{S.S.} = \frac{1.75^2 + 0.20^2 + 0.35^2}{60} - \frac{2.30^2}{180} = 0.0244
$$

Error by difference = 0.3655 with an M.S. of 0.0192.

Thus only B and C are significant, but note the very large Block S.S., much of which is due to II (ii) − II (i), and this would have been included in the error if the partially confounded design had not been used.

To display the results we would build up tables from the effects, so we first convert these to mean effects. The divisors will be the usual ones for the non-confounded effects but, for the confounded ones, having worked out the analysis it is simplest to obtain the divisor as Divisor when unconfounded × amount of information retained. Thus the mean effect BC'

$$= \frac{Q}{18 \times \frac{8}{9}} = \frac{Q}{16} \quad \text{or} \quad \frac{3Q}{48}$$

and its S.E. will be

$$\sqrt{\frac{\text{M.S.}_E \times 2}{16}} = \sqrt{\frac{\text{M.S.}_E}{8}}$$

Likewise mean BC'_{a_0}

$$= \frac{R_0}{6 \times \frac{5}{9}} = \frac{3R_0}{10}$$

and its S.E. will be

$$\sqrt{\frac{3\text{M.S.}_E \times 2}{10}} = \sqrt{\frac{3\text{M.S.}_E}{5}}$$

These are all the basic tools for obtaining confounded designs, and more complicated ones are obtained by combining these ideas. For instance, if we wanted smaller blocks for a $3 \times 3 \times 2$ design calling the factors A, B, C in that order, it is clear that blocks of 6 are required to keep main effects free, and that allows AC and BC to be free as well. We know that we can deal with 3×3 by using the I and J-sets, so can see that we could have a balanced design by this sort of arrangement:

	Blocks					
	I (i)	I (ii)	I (iii)	II (i)	II (ii)	II (iii)
c_0	I_1	I_2	I_3	I_2	I_3	I_1
c_1	I_2	I_3	I_1	I_1	I_2	I_3

and similarly for J.

The analysis is similar to the $3 \times 2 \times 2$. Each I can be estimated by subtracting from the total I-effect the total for blocks in which it occurs, and $\frac{1}{4}$ information is found to be lost. Likewise with the ABC-interaction Ics can be calculated and corrected by subtracting block totals where the I appears with c_1, and adding where it appears with c_0; $\frac{3}{4}$ information is lost.

Partial confounding is also useful when some factors have several levels, but only certain orthogonal contrasts are of interest. For example, when there are two factors each at four levels with hypotheses concerned with linear and quadratic effects, we might well assume that the cubic × cubic interaction did not exist, and be prepared to confound it heavily with blocks to enable small blocks to be used. Here if we set out the coefficients for the cubic × cubic interaction we get

	b_0	b_1	b_2	b_3
a_0	+1	−3	+3	−1
a_1	−3	+9	−9	+3
a_2	+3	−9	+9	−3
a_3	−1	+3	−3	+1

If we make up two blocks of eight, so that one contains only the combinations with negative coefficients and the other the combinations with positive coefficients, any effect of the cubic × cubic interaction will be largely confounded with the difference between the two blocks. We can then use the same method as we did with the $3 \times 2 \times 2$ design to correct the other effects for blocks, and to calculate the extent to which they are confounded. It turns out that all main effects and all interactions involving quadratic effects are unconfounded; but $\frac{1}{25}$th of the information on the linear × linear interaction and $\frac{4}{25}$ths of that on each of the linear × cubic interactions is lost, which together with the loss of $\frac{16}{25}$ths of the cubic × cubic makes unity. This is expected since 1 D.F. must be allocated to the blocks.

CHAPTER TEN

LATTICE DESIGNS AND NON-ORTHOGONAL ANALYSIS

We have now seen how to get small blocks with almost any factorially arranged treatments, but there is another type of treatment which suffers from the same difficulty. Plant breeders are often faced with the problem of comparing a large number of supposedly different genotypes, or they want to do a diallel analysis to discover something about the inheritance of some particular features. This means that they have a large number, perhaps a 100 or more "treatments", and if they are to make useful comparisons must have very low errors of natural variation. Their treatments bear no particular relation to one another, and can seldom be thought of as any factorial combination. Similar problems arise with the testing of drugs. In all these cases the large number of treatments leads to large within-block errors if simple randomized-block designs are used.

We can, however, use a somewhat sophisticated statistical arrangement known as *incomplete-block* or *lattice designs*. These are designs in which the treatments are arranged in randomized incomplete blocks or quasi-Latin squares. There are many of them, and they may be "balanced" when all treatment comparisons have the same error variance, or "partially balanced" where the errors differ slightly. Which to use will depend largely on how many treatments there are, how much experimental material is available, and the physical limits on block size. A balanced design can be had for any number of treatments in any number of units per block, but the number of replications required for balance is fixed by these two numbers, and is often very large. The simplest designs are those for which the number of treatments is a perfect square, and we talk in terms of p^2 treatments; this is no difficulty usually because, if there are say 23 treatments, two treatments would be included twice to bring the number up to 25. Let us consider first the simple case of 9 treatments arranged in randomized incomplete blocks of 3 plots each. These treatments, name them 1–9, could be written in the form of a square lattice as

$$\begin{array}{ccc} 1 & 2 & 3 \\ 4 & 5 & 6 \\ 7 & 8 & 9 \end{array}$$

If in one replicate we divided into blocks according to the rows of this square, in the second replicate according to the columns, in the third according to the I-diagonals, and in the fourth according to the J-diagonals we should have

Replicate I				Replicate II				Replicate III				Replicate IV			
Block				Block				Block				Block			
(1)	1	2	3	(4)	1	4	7	(7)	1	5	9	(10)	1	8	6
(2)	4	5	6	(5)	2	5	8	(8)	4	8	3	(11)	4	2	9
(3)	7	8	9	(6)	3	6	9	(9)	7	2	6	(12)	7	5	3

giving a very balanced arrangement in which each treatment appears with every other treatment once and once only in a block. This leads to relatively simple methods for correcting the treatments for the fact that none appears in every block; for instance, using the ordinary model where b_i and a_i are block and treatment effects respectively, the total for treatment 1 will be

$$A_1 = 4M + 4a_1 + b_1 + b_4 + b_7 + b_{10}$$

whilst the total for blocks 1, 4, 7 and 10, the blocks which contain treatment 1, will be

$$B_1 = 3M + 3b_1 + a_1 + a_2 + a_3$$
$$B_4 = 3M + 3b_4 + a_1 + a_4 + a_7$$
$$B_7 = 3M + 3b_7 + a_1 + a_5 + a_9$$
$$B_{10} = 3M + 3b_{10} + a_1 + a_8 + a_6$$

So the sum of the blocks which contain treatment 1,

$$\sum B_{a_1} = 12M + 3b_1 + 3b_4 + 3b_7 + 3b_{10} + 3a_1 + \sum a$$

(by definition $\sum a = 0$) but from above

$$3A_1 = 12M + 3b_1 + 3b_4 + 3b_7 + 3b_{10} + 12a_1 \quad \text{so} \quad 3A_1 - \sum B_{a_1} = 9a_1$$

We can obtain treatment effects free of block effects by simply correcting treatment totals by the total of the blocks in which they appear.

Now with a slight re-arrangement we can set out four replicates so that each treatment appears with every other treatment in a row and in a column.

Replicate I	Replicate II	Replicate III	Replicate IV
1 2 3	1 4 7	1 5 9	1 8 6
4 5 6	2 5 8	8 3 4	9 4 2
7 8 9	3 6 9	6 7 2	5 3 7

We have produced a *lattice square* with complete balance, in which each treatment appears once with every other treatment in a row and in a column. To do this for p^2 treatments we require $p+1$ replicates and, since p is usually much greater than 3, such designs require too much replication for many purposes; so a compromise of arranging the treatments so that each appears with every other once in a row *or* in a column, which requires $\frac{1}{2}(p+1)$ replicates but can only be got when p is an odd number, is more commonly used. It deals very well with 25 treatments in 3 replicates, 49 in 4, and 81 in 5. We will consider as an example the case of 25 selections from a cross of winter wheat varieties using this partial balance.

The procedure is first to choose the squares. They can be worked out, or many useful ones are given by Cochran and Cox (1957). For 25 treatments suitable squares can be obtained by taking the numbers in natural order for the first; then for the second square write 1 first, and go down the rows of the first square moving two places to the right, i.e. 1, 8, 15, 17, 24. Start the next row with the next number (25) and continue. The third square is generated in the same way, but moving four places to the right each time.

1	2	3	4	5		1	8	15	17	24		1	10	14	18	22
6	7	8	9	10		25	2	9	11	18		23	2	6	15	19
11	12	13	14	15		19	21	3	10	12		20	24	3	7	11
16	17	18	19	20		13	20	22	4	6		12	16	25	4	8
21	22	23	24	25		7	14	16	23	5		9	13	17	21	5

Rows and columns are then randomized separately and independently for each square, and the position of the squares randomized in the field. The treatments are randomized to the numbers 1–25.

When the results have been obtained, they are set out on a plan of the layout as shown in Table 10.1.

Table 10.1—Layout and analysis of a lattice square experiment

Layout and yields

Square I

										Total (R)	L	δ
(18)	33·4	(11)	35·6	(2)	30·4	(9)	31·1	(25)	30·1	160·6	−33·4	−2·7
(12)	28·7	(10)	28·7	(21)	25·4	(3)	24·5	(19)	35·2	142·5	+22·4	+1·8
(24)	24·9	(17)	29·2	(8)	35·3	(15)	30·9	(1)	32·7	153·0	−11·5	−0·9
(6)	27·1	(4)	26·1	(20)	25·1	(22)	27·4	(13)	29·6	135·3	+11·7	+0·9
(5)	40·6	(23)	30·2	(14)	30·5	(16)	36·0	(7)	33·9	171·2	−67·0	−5·4
Total (C)	154·7		149·8		146·7		149·9		161·5	762·6		
M	−17·0		−12·8		−10·8		+3·8		−41·0		−77·8	
ε	−0·9		−0·6		−0·5		+0·2		−2·1			

Square II

(1)	28·8	(3)	32·5	(5)	32·3	(4)	32·2	(2)	32·5	158·3	−31·2	−2·5
(16)	27·9	(18)	34·7	(20)	31·3	(19)	39·3	(17)	33·4	166·6	−47·4	−3·8
(11)	27·6	(13)	22·0	(15)	37·7	(14)	28·0	(12)	31·4	146·7	−5·9	−0·5
(21)	25·1	(23)	23·2	(25)	32·8	(24)	28·7	(22)	43·3	153·1	−47·2	−3·8
(6)	34·5	(8)	34·4	(10)	40·0	(9)	32·2	(7)	37·7	178·8	−68·8	−5·5
Total (C)	143·9		146·8		174·1		160·4		178·3	803·5		
M	−12·6		−11·7		−62·7		−29·6		−83·9		−200·5	
ε	−0·6		−0·6		−3·1		−1·5		−4·2			

Square III

(21)	27·6	(17)	28·3	(5)	30·5	(13)	28·0	(9)	31·2	145·6	+9·7	+0·8
(15)	26·5	(6)	22·2	(19)	29·1	(2)	26·5	(23)	21·8	126·1	+68·8	+5·5
(7)	17·6	(3)	23·0	(11)	19·9	(24)	24·3	(20)	17·1	101·9	+98·0	+7·8
(18)	21·5	(14)	25·5	(22)	18·4	(10)	27·1	(1)	17·8	110·3	+106·9	+8·6
(4)	33·3	(25)	28·9	(8)	34·6	(16)	30·9	(12)	32·3	160·0	−5·1	−0·4
Total (C)	126·5		127·9		132·5		136·8		120·2	643·9		
M	+64·1		+46·8		+86·0		+27·1		+54·3		+278·3	
ε	+3·2		+2·3		+4·3		+1·4		+2·7			

Treatment number is shown in brackets in each cell. Yields are expressed in lb per plot.

Treatment totals (unadjusted) (A)

(1)	79·3	(2)	89·4	(3)	80·0	(4)	91·6	(5)	103·4
(6)	83·8	(7)	89·2	(8)	104·3	(9)	94·5	(10)	95·8
(11)	83·1	(12)	92·4	(13)	79·6	(14)	84·0	(15)	95·1
(16)	94·8	(17)	90·9	(18)	89·6	(19)	103·6	(20)	73·5
(21)	78·1	(22)	89·1	(23)	75·2	(24)	77·9	(25)	91·8

Grand total (G) 2210·0

Analysis of variance

Source of variation	D.F.	S.S.	M.S.
Squares	2	549·80	
Treatments (unadjusted)	24	615·43	
Rows within squares (adjusted)	12	591·21	49·27
Columns within squares (adjusted)	12	234·82	19·57
Error	24	235·59	9·82
Total	74	2226·85	

LATTICE DESIGNS AND NON-ORTHOGONAL ANALYSIS

Treatment totals (adjusted)

(1)	84·5	(2)	86·4	(3)	89·0	(4)	90·7	(5)	96·6
(6)	85·5	(7)	83·0	(8)	100·7	(9)	88·5	(10)	98·4
(11)	90·8	(12)	90·9	(13)	79·5	(14)	87·0	(15)	99·5
(16)	86·2	(17)	84·5	(18)	93·4	(19)	107·8	(20)	77·5
(21)	79·0	(22)	95·1	(23)	73·0	(24)	80·0	(25)	82·0

To analyse we first obtain totals for each row (R), each column (C), and each treatment (A), checking carefully to see that the sum of the rows equals the sum of the columns in each square (S), and that the sum of the squares equals the sum of the treatment totals (G).

The analysis of variance is shown in Table 10.1.

Sum of squares for squares is obtained in the usual way, i.e.

$$\frac{\sum S^2}{p^2} - \frac{G^2}{rp^2} = \frac{762 \cdot 6^2 + 803 \cdot 5^2 + 643 \cdot 9^2}{25} - \frac{2210 \cdot 0^2}{75} = 549 \cdot 80$$

and the total S.S. can be got as usual as

$$\sum x^2 - \frac{(\sum x)^2}{rp^2} = 2226 \cdot 85$$

It is a feature of most experiments that need one component in the analysis to be adjusted for the effect of another component, that

the S.S. of say A unadjusted + S.S. of B adjusted

= the S.S. of A adjusted + S.S. of B unadjusted

and in the analysis-of-variance table we can work out one of treatments, rows or columns in the unadjusted form, and in this design we do so with treatments, so their sum of squares will be

$$\frac{\sum A^2}{r} - \frac{G^2}{rp^2} = \frac{79 \cdot 3^2 + 89 \cdot 4^2 + \ldots + 91 \cdot 8^2}{3} - \frac{2210 \cdot 0^2}{75} = 615 \cdot 43$$

Now the rows and columns must be adjusted for the treatments they contain before we can get their sums of squares. This is most simply done by writing down values called L and M such that L_i = total from all replications of all treatments in the ith row minus $r \times$ the total of the ith row. Thus the first row of Square II contains treatments 1, 2, 3, 4 and 5, so

$$L_6 = A_1 + A_2 + A_3 + A_4 + A_5 - 3R_6$$
$$= 79 \cdot 3 + 89 \cdot 4 + 80 \cdot 0 + 91 \cdot 6 + 103 \cdot 4 - 3 \times 158 \cdot 3 = -31 \cdot 2$$

Similarly M_i = the total from all replicates of all treatments included in the ith column minus $r \times$ the total of the ith column. When we have written down all the Ls and Ms, we add them for each square to get values we will call L_r and M_r, and $L_r = M_r$ for each square, and as a further check L_r for any square = Grand total $- r \times$ the square total. One further check is that $\sum L_r = 0$.

The sum of squares for rows adjusted for treatments is now simply

$$\frac{\sum L^2}{pr(r-1)} - \frac{\sum L_r^2}{p^2 r(r-1)}$$

$$= \frac{(-33\cdot4)^2 + 22\cdot4^2 + \ldots + (-5\cdot1)^2}{30} - \frac{(-77\cdot8)^2 + (-200\cdot5)^2 + 278\cdot3^2}{150}$$

$$= 591\cdot21$$

Similarly the sum of squares for columns adjusted is

$$\frac{\sum M^2}{pr(r-1)} - \frac{\sum M_r^2}{p^2 r(r-1)} = 234\cdot82$$

We now get the sum of squares for error by difference.

Before any treatment comparisons can be made, we have to adjust the treatment totals for the effects of the rows and columns in which they appear. We work out the mean squares for rows, columns and error.

Let these be E_r, E_c and E_e respectively, then calculate

$$\lambda = \frac{E_r - E_e}{p(r-1)E_r} = \frac{49\cdot27 - 9\cdot82}{5 \times 2 \times 49\cdot27} = 0\cdot080$$

and

$$\mu = \frac{E_c - E_e}{p(r-1)E_c} = \frac{19\cdot57 - 9\cdot81}{5 \times 2 \times 19\cdot57} = 0\cdot050$$

If E_r is less than E_e, there is obviously no row effect, and so no adjustment for row effects is needed, and similarly if E_c is less than E_e; but if λ and μ are positive quantities, we now weight the row and column effects according to their contributions to the variance by multiplying all the Ls by λ (call these values δ) and all the Ms by μ calling these ε.

The adjusted total for any treatment is now obtained by adding to the unadjusted total the δ and ε-values for each row and column in which the treatment appears. Thus for treatment 1 we have

$$A_1 + \delta_3 + \delta_6 + \delta_{14} + \varepsilon_5 + \varepsilon_6 + \varepsilon_{15}$$

The approximate S.E. for an adjusted treatment total is

$$\sqrt{rE_e\left\{1+\frac{rp}{p+1}(\lambda+\mu)\right\}} = \sqrt{3\times 9\cdot 82\left\{1+\frac{2\times 5}{6}(0\cdot 080+0\cdot 050)\right\}} = 5\cdot 99$$

This is usually good enough to do a Duncan's Multiple Range test or for similar purposes. More accurate S.E.s can be calculated for particular comparisons (see Cochran and Cox, 1957, p. 488).

Generally the F-test is not very helpful in explaining the problem being investigated, and it is typical of these designs that they do not furnish an exact F-test. However, an approximate one can be obtained by getting the sum of squares of the adjusted treatment totals and comparing the mean square with the effective error:

$$E_e\left\{1+\frac{rp}{p+1}(\lambda+\mu)\right\}$$

Another partially balanced design useful for biologists is one which is applicable when there is the possibility of using the same experimental material for more than one treatment, and when treatments can be applied in sequence to the same material. The idea was first used in research with dairy cows when there were feeding treatments which were effective in a short time, and where the effect of the treatment, as measured by the milk production, would die out in a short time after feeding with that treatment ceased. Now, even if we used a single cow as an experimental unit in an ordinary completely randomized experiment, we should require a large number of cows because the genetic variation and variation due to stage of lactation, which would be the main contributors to the error of such an experiment, would be very large; so the chances of showing up small differences are very remote unless something is done to control this variation. The first thought is to try all the treatments on each cow to remove the genetic variation, i.e. a randomized-block design in which a cow would be a block. But since the yield of a particular cow diminishes as the lactation proceeds, treatments which came early in the lactation would have an advantage over those that came later. Hence a Latin square would be better, where columns might be allocated to cows and rows to period of time, e.g.

Cows	I	II	III	IV
1st fortnight	A	B	C	D
2nd fortnight	B	C	D	A
3rd fortnight	C	D	A	B
4th fortnight	D	A	B	C

Now each treatment is used on every cow, and each treatment is present in every period. Many useful experiments were made in this simple way, but it soon became apparent that experiments where a single object, such as a cow, receives different treatments in sequence were open to criticism not inherent in a Latin square as used in field experiments. Looking at the particular square, it is seen that, except in the first period, whenever A appears, that particular cow has been having D in the previous period. Similarly, C always follows B, and so on. If we compare the yields when A is given with those when C is given, we cannot be sure we are comparing A with C, because we might equally well be comparing the after-effects or residual effects of D with those of B. To overcome this we need a special arrangement of a Latin square in which both direct and residual effects can be calculated, and we get the best results and easiest computing if the direct effects are balanced for residual effects, i.e. each treatment follows every other treatment the same number of times. This is easily achieved for squares of even numbers by writing the first column in the order 1st letter, 2nd letter, last letter, 3rd letter, next to last letter, and so on, and then filling in the rows in natural order, e.g. with 6 treatments, we have

Period						
1	A	B	C	D	E	F
2	B	C	D	E	F	A
3	F	A	B	C	D	E
4	C	D	E	F	A	B
5	E	F	A	B	C	D
6	D	E	F	A	B	C

when it is seen that A follows B in the 6th period, C in 3rd, D in 4th, E in 5th, F in 2nd, and likewise all letters follow every other letter once and once only. Therefore when we get a treatment total it will include six times its own direct effect, and the residual effect of each of the other treatments; it is not difficult to correct for the residual effects by a system somewhat similar to that used for incomplete-block designs. Furthermore, if we sum all yields in periods following a particular treatment, such totals must contain five times the residual effect of that treatment, plus the direct effect of each of the other treatments, and so the residual effects themselves can be fairly easily calculated.

If we had an odd number of treatments, we cannot have this arrangement in a single square, but we can in two squares, so we write the first column of the first square as with even numbers, and write the first column of the second square as the reverse of this, i.e. 5 would have ABECD and DCEBA.

Table 10.2—An experiment with treatments applied in sequence

Treatments

a_1 25% ⎫ a_4 25% ⎫
a_2 40% ⎬ short straw a_5 40% ⎬ long straw
a_3 55% ⎭ a_6 55% ⎭

Layout and food intake (kg)

Animal

Period (time)	1	2	3	4	5	6	Total
1	a_6 222	a_1 272	a_5 287	a_2 260	a_3 261	a_4 309	1611 P_1
2	a_2 284	a_4 265	a_1 274	a_3 271	a_6 268	a_5 227	1589 P_2
3	a_4 294	a_6 261	a_3 240	a_5 339	a_1 233	a_2 294	1661 P_3
4	a_3 212	a_5 259	a_4 293	a_6 302	a_2 279	a_1 238	1583 P_4
5	a_1 204	a_3 256	a_2 236	a_4 337	a_5 260	a_6 228	1521 P_5
6	a_5 230	a_2 260	a_6 188	a_1 246	a_4 214	a_3 217	1355 P_6
Total	1446	1573	1518	1755	1515	1513	9320
	F_5	F_2	F_6	F_1	F_4	F_3	G

Treatments	Total (A)	R	$\hat{A}$	$\hat{R}$	$\hat{R}'$
a_1	1467	1242	2550	−850	−332
a_2	1613	1251	1547	−742	−1100
a_3	1457	1364	−2359	2030	2608
a_4	1712	1182	3946	−2980	−3932
a_5	1602	1340	1635	1634	1342
a_6	1469	1330	−2210	908	1414

Correction factor = 2 412 844

Analysis of variance

Source of variation	D.F.	S.S.	M.S.
Animals	5	9487	
Periods	5	9575	
Direct effects (unadjusted)	5	9158	
Residual effects (adjusted)	5	2938	588
Residual effects (unadjusted)	5	4371	
Direct effects (adjusted)	5	7725	1545
Error	15	12 200	813
Total	35	43 358	

Adjusted effects

	Direct	Residual
a_1	244	254
a_2	268	254
a_3	245	271
a_4	282	241
a_5	269	269
a_6	246	264
S.E.	±11·8	±13·2

In all cases columns are randomized to cows, or whatever material is used, and treatments are randomized to the letters; but rows must not be randomized, as the sequences must stay as they are.

Table 10.2 shows an example in which the effect of type and quantity of roughage on the voluntary food intake of cattle was studied. The treatments were really a factorial arrangement of short vs. long straw, each at three levels of inclusion in the diet, but for the sake of the preliminary analysis we may consider them as a single factor (A) at six levels.

To analyse we first obtain row, column, and treatment totals in the usual way. Sums of squares for rows, columns and total are obtained as usual and entered in the analysis-of-variance table. As stated earlier,

the S.S. of direct effects adjusted for residual effects
+ S.S. for residual effects not adjusted
= S.S. for direct not adjusted + S.S. for residual adjusted

so the treatment effects have only $2(n-1)$ not $4(n-1)$ D.F. altogether for an $n \times n$ square. We need only the adjusted for testing, but we need one of the unadjusted for getting the error by difference.

Sum of squares for direct effects unadjusted is the usual $(\sum A^2)/n -$ C.F. The others are more complex. First we add together all the yields following a particular treatment and call these R (for a_1 this is the intake in period 6 for animal 1 + that in period 2 for animal 2, etc.).

Then write down F for each treatment where $F =$ the total of the columns in which this treatment comes last (for treatment a_1 this is column 4). (If we had more than one square, we should add appropriate columns together; with one square it is just a single-column total.)

Calling the period totals $P_1, \ldots, P_6$, and the grand total G, we can work out the direct effect for each treatment as

$$a_i = \frac{(n^2 - n - 1)A_i + nR_i + F_i + P_1 - nG}{mn(n^2 - n - 2)}$$

where m is the number of $n \times n$ squares in the experiment. In our case m is 1 and

$$a_1 = \frac{29A_1 + 6R_1 + F_1 + P_1 - 6G}{168}$$

This formula can be obtained by simple, though tedious, algebra but we can convince ourselves by applying it in this particular case. Calling the general mean M, the direct effects a_i, the residual effects r_i, and the period and animal effects p_i and c_i respectively, we have

$$A_1 = 6M + 6a_1 + \sum r - r_1 + \sum p + \sum c$$
$$R_1 = 5M + 5r_1 + \sum a - a_1 + \sum p - p_1 + \sum c - c_4$$
$$F_1 = 6M + \sum a + \sum r - r_1 + \sum p + 6c_4$$
$$P_1 = 6M + \sum a + 6p_1 + \sum c$$
$$G = 36M + 6\sum a + 5\sum r + 6\sum p + 6\sum c$$

Remembering that by definition, $\sum a = \sum r = \sum p = \sum c = 0$

$$29A_1 = \quad 174M + 174a_1 - 29r_1$$
$$6R_1 = \quad 30M - \quad 6a_1 + 30r_1 - 6p_1 - 6c_4$$
$$F_1 = \quad 6M \quad\quad - 1r_1 \quad\quad + 6c_4$$
$$P_1 = \quad 6M \quad\quad\quad\quad + 6p_1$$
$$-6G = -216M$$

Thus on adding, $29A_1 + 6R_1 + F_1 + P_1 - 6G = 168a_1$.

To obtain an adjusted direct-effect mean, we simply add a to the general mean, but to obtain the sum of squares for adjusted means there is no need to divide by the 168.

Write down a column of $\hat{A}s = 168a$ and the

$$\text{S.S.} = \frac{\sum \hat{A}^2}{mn(n^2-n-1)(n^2-n-2)} = \frac{37\,634\,812}{6 \times 29 \times 28} = 7725$$

(Note there is no correction factor because the $\hat{A}$ sum to zero.) Likewise we can calculate adjusted residual effects for each treatment

$$\hat{R}_i = mn(n^2 - n - 2)r_i = nA_i + n^2 R_i + nF_i + nP_1 - (n+2)G$$

i.e.

$$168r_i = 6A_i + 36R_i + 6F_i + 6P_1 - 8G$$

in the present case, when residual-effect means can be got by adding r to the general mean, and the S.S. for residual effects adjusted is

$$\frac{\sum \hat{R}^2}{mn^3(n^2-n-2)} = \frac{17\,768\,784}{6 \times 6 \times 6 \times 28} = 2938$$

To get the S.S. for residual effects unadjusted, as a check we need another

set of quantities
$$\hat{R}'_i = \hat{R}_i + G - nA_i = \hat{R}_i + G - 6A_i$$
in our case.

Then the S.S.
$$= \frac{\sum \hat{R}'^2}{mn^3(n^2-n-1)} = \frac{27\,382\,872}{6 \times 6 \times 6 \times 29} = 4371$$

The error S.S. is got by subtracting from the total S.S. the animals and periods S.S., and the S.S. of either of the two sets of treatment components (not both). The adjusted mean squares can provide an F-test and, if standard errors are required, that for comparing direct-effect means is

$$\sqrt{\left(\frac{\text{M.S.}_E}{mn} \times \frac{(n^2-n-1)}{(n^2-n-2)}\right)} = \sqrt{\left(\frac{813}{6} \times \frac{29}{28}\right)} = \pm 11 \cdot 8$$

and that for comparing residual effect means is

$$\sqrt{\left(\frac{\text{M.S.}_E}{mn} \cdot \frac{n^2}{n^2-n-2}\right)} = \sqrt{\left(\frac{813}{6} \times \frac{36}{28}\right)} = \pm 13 \cdot 2$$

In the present case these S.E.s might be adequate to test the null hypotheses that led to the factorial arrangement of treatments. Setting the corrected means out in two-way tables, with marginal means and S.E.s, we have

Quantity of roughage (%)	Direct effects				Residual effects			
	Straw				Straw			
	Short	Long	Mean		Short	Long	Mean	
25	244	282	263		254	241	248	
40	268	269	268	($\pm 8\cdot 3$)	254	269	262	($\pm 9\cdot 3$)
55	245	246	246		271	264	268	
	($\pm 11\cdot 8$)				($\pm 13\cdot 2$)			
Mean	252	266			260	258		
	($\pm 6\cdot 8$)				($\pm 7\cdot 6$)			

In fact in this experiment, which was based on much previous investigation, the null hypotheses were not the usual ones for a factorial arrangement but were: there is no linear effect of quantity of roughage, and there are no deviations from linear effect when short straw is fed; there is no linear effect and there are no deviations from linear effect when long straw is fed; and there is no difference in intake on average due to length of straw. For direct effects, the difference observed for the last of these is $266 - 252 = 14$ with an S.E. of $6\cdot 8 \times \sqrt{2} = 9\cdot 6$, so with t of less than 2 we have little reason to doubt the null hypothesis. It is obvious that there is no linear effect of quantity for short straw, and the deviations from linear are $244 + 245 - 2 \times 268 = -47$ with an S.E. of $11 \cdot 8 \sqrt{(1^2 + 1^2 + 2^2)} = 28 \cdot 9$, giving $t = 1 \cdot 63$, again not strong evidence against the null hypothesis. For long straw, the linear effect is -36, with an S.E. of $16 \cdot 7$ giving $t = 2 \cdot 16$, which achieves the 5 per cent level of significance, whilst the deviations are $-10 \pm 28 \cdot 9$ and so are obviously not significant.

The table of residual effects indicates that none of the effects to be tested can possibly achieve the 5 per cent level of significance; so all we could conclude from the experiment would be that, when feeding long straw, there is an immediate linear decline in voluntary intake of cattle as the proportion of straw in the diet is increased within the range of 25–55 per cent. The experimenter also noticed that there was a large difference in intake between short- and long-straw diets at the 25 per cent level of inclusion, which was rather surprising at the time; and this led to further experiments to compare various ways of treating the straw. It would have been very unwise to change hypotheses at the end to justify "testing" this in this experiment.

It should be stressed that this type of analysis can only be used when the treatments leave the experimental organism functioning normally, and the periods must not be so long that the organism undergoes large changes in physiological state during the course of the experiment. It should also be realized that only a very small sample of the population of, say, cows is explored in a single experiment, so that extrapolation to cows in general may be hazardous. This is particularly the case where surgical operations have to be performed in order to obtain the data, e.g. when the data are quantities of volatile fatty acids in the rumen obtained from rumen fistulae. Then the animals are usually specially selected for all sorts of "practical" reasons, rather than selected at random from a known population, and it takes a considerable act of faith to believe that results from this artificial population of animals with holes in their sides will have anything to do with the effects on normal animals of that species kept under normal conditions.

Consideration of these non-orthogonal analyses prompts the question: can we not analyse any design, and therefore why concern ourselves to start with balanced designs? There is no doubt that we can get some analysis from every experiment with a reasonable number of observations, no matter what its design, but there are problems. In the days before electronic computers there were great arithmetical problems, in that the calculations were very arduous, so biologists seldom attempted to do it. Now with package computer programs the greater problem is the biological one of making sensible assumptions in the mathematical models required. Basically we can always assume that a particular observation $x_{ijk...z}$ will have a value

$$x_{ijk...z} = M + a_i + b_j + c_k \ldots r_z + \varepsilon_{ijk...z}$$

The method of analysis which is equivalent to all those shown previously, and indeed on which they are all based, is known as the *method of least squares*, which simply means that $M, a, b, c, \ldots, r$ are estimated so that $\sum \varepsilon_{ijk...z}^2$ is a minimum. To do this we write an equation in the form

$$\sum \varepsilon_{ijk...z}^2 = \sum (x_{ijk...z} - M - a_i - b_j - \ldots - r_z)^2$$

and differentiate this sum of squares with respect to each unknown in turn and set the derivatives equal to zero. The resulting equations are known as *normal* equations. Given sufficient equations the parameters can be calculated by ordinary algebraic methods, but in addition to the *normal* equations, we have to use constraints or make assumptions about the parameters.

Mathematically several different sets of constraints may be possible but, unless that chosen is biologically meaningful, no useful analysis will be obtained. So the biologist needs to decide upon these constraints before beginning the experiment, and needs to put them into terms which can be used in conjunction with the model.

As an example, consider an experiment to test the difference in live-weight gain of lambs between two breeds of sheep, when we are also interested in the effect of parity (number of lambs per birth). Suppose we start with nine ewes of each breed, chosen at random, and obtain the following results (the first value in each cell is the number of lambs obtained, and the second the average live-weight gain (kg) of those lambs):

		Litter size (l)		
		1	2	3
Breed (b)	1	3, 20	8, 20	6, 18
	2	4, 20	6, 15	6, 10

If this was considered as a balanced design, the mean live-weight gain for Breed 1 would be

$$\frac{20 + 20 + 18}{3} = 19 \cdot 33$$

and for Breed 2
$$\frac{20+15+10}{3} = 15\cdot00$$

The biologist would know that this would not make a very useful comparison, because it assumes that the number of singles, twins and triplets is equal and the same in each breed, and he has no reason to believe that that is true. Indeed his best evidence is that the ratio is 3:8:6 in Breed 1 and 4:6:6 in Breed 2; so he would prefer to assume that this represents the usual distribution of births and compare the weighted means. Then the mean for Breed 1 is

$$\frac{3\times 20+8\times 20+6\times 18}{3+8+6} = 19\cdot29$$

(very similar to the previous estimate because there is little difference in live weight between the three litter sizes) but that for Breed 2 is

$$\frac{4\times 20+6\times 15+6\times 10}{16} = 14\cdot38$$

and the difference between the breeds is now 4·91 instead of 4·33 as estimated previously. Likewise we should wish to have weighted averages for litter-size effects.

Now, for analysis, the basic model for the kth animal of the ith breed born as one of litter size j is:

$$x_{ijk} = M + b_i + l_j + (bl)_{ij} + \varepsilon_{ijk}$$

where M is a constant effect, b_i is the effect due to the ith breed, l_j is the effect due to the jth litter size, $(bl)_{ij}$ is the additional effect of the ith breed with the jth litter size and ε_{ijk} the natural variation.

In the balanced case we should estimate b_i as the difference between the mean of all the animals of the ith breed and the general mean, and similarly obtain estimates of l_j and $(bl)_{ij}$; we could easily write down the constraints (which are really part of the definitions of the terms in the equation). There are two breeds, so i can take the value 1 or 2; there are three litter sizes, so j can be 1, 2 or 3, and the constraints are

$$b_1+b_2 = 0$$
$$l_1+l_2+l_3 = 0$$
$$(bl)_{11}+(bl)_{12}+(bl)_{13} = 0$$
$$(bl)_{21}+(bl)_{22}+(bl)_{23} = 0$$
$$(bl)_{11}+(bl)_{21} = 0$$
$$(bl)_{12}+(bl)_{22} = 0$$
$$(bl)_{13}+(bl)_{23} = 0$$

although this last one is redundant since the four equations above it ensure that it is so.

In the present case we have decided for biological reasons that such estimates of the bs and ls are not meaningful, so we must re-define them to give the effects we are seeking. Using the standard mathematical notation, where μ_{ij} stands for the mean live-weight gain of the lambs of the ith breed born in the jth litter size, $\bar{\mu}_{i.}$ the mean (to be defined for biological reasons) of lambs of the ith breed, $\bar{\mu}_{.j}$ likewise the mean of all lambs born in the jth litter size, $\bar{\mu}_{..}$ stands for the mean of the values of $\bar{\mu}_{i.}$ and equally the mean of the $\bar{\mu}_{.j}$, we can set out our table of numbers of animals and means as

		Litter size			Mean
		1	2	3	
Breeds	1	3, μ_{11}	8, μ_{12}	6, μ_{13}	$\bar{\mu}_{1.}$
	2	4, μ_{21}	6, μ_{22}	6, μ_{23}	$\bar{\mu}_{2.}$
Mean		$\bar{\mu}_{.1}$	$\bar{\mu}_{.2}$	$\bar{\mu}_{.3}$	$\bar{\mu}_{..}$

We considered that the most appropriate means for breeds were those obtained by weighting the individual means according to the number of lambs making up each individual mean, so

$$\bar{\mu}_{1.} = \frac{3\mu_{11} + 8\mu_{12} + 6\mu_{13}}{3 + 8 + 6} \quad \text{and} \quad \bar{\mu}_{2.} = \frac{4\mu_{21} + 6\mu_{22} + 6\mu_{23}}{4 + 6 + 6}$$

Likewise

$$\bar{\mu}_{.1} = \frac{3\mu_{11} + 4\mu_{21}}{3 + 4} \quad \bar{\mu}_{.2} = \frac{8\mu_{12} + 6\mu_{22}}{8 + 6} \quad \bar{\mu}_{.3} = \frac{6\mu_{13} + 6\mu_{23}}{6 + 6}$$

and

$$\bar{\mu}_{..} = \frac{3\mu_{11} + 8\mu_{12} + 6\mu_{13} + 4\mu_{21} + 6\mu_{22} + 6\mu_{23}}{3 + 8 + 6 + 4 + 6 + 6}$$

Now we can define the M, bs, ls and (bl)s in terms of the $\bar{\mu}$s.

$M = \bar{\mu}_{..}$, our best estimate of the live-weight gain of lamb
$b_1 = \bar{\mu}_{1.} - \bar{\mu}_{..}$ $b_2 = \bar{\mu}_{2.} - \bar{\mu}_{..}$
$l_1 = \bar{\mu}_{.1} - \bar{\mu}_{..}$ $l_2 = \bar{\mu}_{.2} - \bar{\mu}_{..}$
                          $l_3 = \bar{\mu}_{.3} - \bar{\mu}_{..}$
$(bl)_{11} = \mu_{11} - \bar{\mu}_{1.} - \bar{\mu}_{.1} + \bar{\mu}_{..}$ $(bl)_{12} = \mu_{12} - \bar{\mu}_{1.} - \bar{\mu}_{.2} + \bar{\mu}_{..}$

and so on, in general,

$$(bl)_{ij} = \mu_{ij} - \bar{\mu}_{i.} - \bar{\mu}_{.j} + \bar{\mu}_{..}$$

The constraints are then obtained algebraically and no more biological decision is required. In the present case, calling the number of animals of the ith breed born in the jth litter size n_{ij}, and taking $n_{i.}$ and $n_{.j}$ as the total number of animals for the ith breed and jth litter size respectively, we have (P. M. Lerman, personal communication)

$$n_{1.}b_1 + n_{2.}b_2 = 0$$
$$n_{.1}l_1 + n_{.2}l_2 + n_{.3}l_3 = 0$$
$$n_{11}\{(bl)_{11}+l_1\} + n_{12}\{(bl)_{12}+l_2\} + n_{13}\{(bl)_{13}+l_3\} = 0$$
$$n_{21}\{(bl)_{21}+l_1\} + n_{22}\{(bl)_{22}+l_2\} + n_{23}\{(bl)_{23}+l_3\} = 0$$
$$n_{11}\{(bl)_{11}+b_1\} + n_{21}\{(bl)_{21}+b_2\} = 0$$
$$n_{12}\{(bl)_{12}+b_1\} + n_{22}\{(bl)_{22}+b_2\} = 0$$
$$n_{13}\{(bl)_{13}+b_1\} + n_{23}\{(bl)_{23}+b_2\} = 0$$

Taking these together with the *normal* equations, we can apply the standard methods for solving simultaneous equations which are available on many electronic computers, and obtain best estimates of all the desired effects, and an estimate of the residual or error variation. However, before starting the experiment, it is wise to inquire how many parameters can be estimated by the computer that it is intended to use, since it requires great capacity, even for an experiment with, say, three factors, each at three levels. (There are 64 equations, which would be considered quite a small experiment if a balanced design were used.) Also, as the number of parameters gets nearer the limit of the capacity, they are estimated with less accuracy, because the rounding errors become serious. Thus the best advice is to use balanced designs which give orthogonal analysis, unless there is some overriding biological reason for not doing so, as in the example with the sheep, where the biology of the sheep themselves dictates the balance of the design. It can be seen that the same type of argument could have applied to the choice of number of ewes in the experiment, which was entirely in the hands of the experimenter. By taking equal numbers of the two breeds, his comparison of, say, singles and twins applies to situations where the two breeds are present in equal quantities; but if he wished to assess this difference for an area where these breeds were present in known unequal numbers, he would require his means weighted for number of each breed in the population. Again this can be done by using the non-orthogonal method of analysis.

One further use is to overcome difficulties which arise during an experiment. For example, in an experiment with eight treatments in four randomized blocks investigating the effect of various potassium treatments on winter wheat, an experimenter needed to take samples of the upper two leaves and ears on a particular day, and managed to get the services of five samplers. Thus he could not conveniently make full use of his samplers by allocating them so that any differences between them would be confounded with blocks, so he allowed them to work across the experiment, assuming that since the treatments came in random order, each man had the same chance as any other man of sampling any particular treatment. At the end of the exercise he discovered that one sampler had misinterpreted the instructions, and the results from his samples were likely to differ from the others. Fortunately each man's samples were recognizable, and non-orthogonal analysis could be made by inserting extra parameters into the model so that, if $x_{ij(k)}$ is the value for the plot having the ith treatment (a_i) in the jth block (b_j) and sampled by the kth sampler (s_k),

$$x_{ij(k)} = M + a_i + b_j + s_k + \varepsilon_{ij}$$

In this case the experimenter has to decide first if the effect of sampler can be simply to add some quantity to the result, because only if this is so is it appropriate to include the term $+s_k$. Then he must decide if there are likely to be any interactions, e.g. between sampler and treatment, and, if he thought this likely, terms such as $(as)_{ik}$ must be included. This emphasizes once more that it is the biologist's responsibility to make the model conform with the biological situation.

Non-orthogonal analyses often become necessary with animal experiments when several animals die during the course of the experiment, though in the simpler cases approximations can be made by estimating the missing values and proceeding with an orthogonal analysis. This is based on a proposal of Yates (1933) to insert a value which minimizes the error S.S. In a single-factor design, the missing value is simply ignored and S.S. and D.F. calculated from the values that are present. In a randomized-block design with r blocks and p treatments, we find the total for the block in which the value is missing (B), the total for the treatment in which the value is missing (A), and the grand total of all the values we have (G), and estimate the missing value as

$$x = \frac{rB + pA - G}{(r-1)(p-1)}$$

Since there are only $(rp-1)$ observations, the total S.S. has only $(rp-2)$ D.F. Consequently, since we still have complete estimates of block and treatment effects with $(r-1)$ and $(p-1)$ D.F. respectively, the error S.S. will have only $\{(r-1)(p-1)-1\}$ D.F.; in other words we must subtract one from the error degrees of freedom. An F-test will give a value of F that is slightly too large, but this is not serious. If a t-test involving a comparison of the mean of the treatment containing the inserted value is made, account should be taken of the fact that that mean is based on only $(r-1)$ replications. Cochran and Cox (1957, p. 111) give an approximate S.E. of the difference between the mean of the treatment with a missing value and any other treatment as

$$\sqrt{\left[\text{M.S.}_E\left(\frac{2}{r}+\frac{p}{r(r-1)(p-1)}\right)\right]}$$

If two values are missing, one should first be guessed, and the remaining one calculated by the formula. Then, using this value, calculate the second one. With this approximation, recalculate the first, and so on, until the new approximations do not differ from the previous ones within the number of decimal places chosen for the data. One degree of freedom is subtracted from the error D.F. for each missing value.

Similarly for an $r \times r$ Latin square, we obtain the totals for the row (R), column (C), and treatment (A) in which the value is missing, and the missing value is inserted as

$$x = \frac{r(R+C+A)-2G}{(r-1)(r-2)}$$

Again one is subtracted from the error D.F. and Cochran and Cox's (1957, p. 126) S.E. for the difference between the mean of the treatment with the missing value and any other treatment mean is

$$\sqrt{\left[\text{M.S.}_E\left(\frac{2}{r}+\frac{1}{(r-1)(r-2)}\right)\right]}$$

For a split-plot design with one value missing, Anderson (1946) gives a formula for the missing value:

$$x = \frac{rM+qT-A}{(r-1)(q-1)}$$

where M is the total of the main plot with the missing value, T the treatment total, A the total of the main treatment with the value missing, r is the number of replications, and q the number of sub-treatments.

Formulae are also available for most useful designs involving confounding of interactions with blocks and for lattice designs; several are given in Cochran and Cox (1957).

CHAPTER ELEVEN

ANALYSIS OF COVARIANCE

There is another method of controlling or reducing the error variation in an experiment, which follows from the idea of the model discussed in Chapter 10, and makes use of the procedures of regression outlined in Chapter 7. If we can obtain a measure of each item before the experiment starts, which is closely related to the final measurement, we can use this initial measurement to adjust the final measurements, thereby reducing the error variation. Or in terms of a model of a single-factor design, if the final measure is y, the initial measure x, and M, a and ε defined as before, we have

$$y_{ij} = M + a_i + b(x_{ij} - \bar{x}) + \varepsilon_{ij}$$

where b is the regression of y on x after allowing for the treatment effect. This term can be put into any other model, e.g. a randomized-block design, but it should be noticed that it is assumed that it is independent of treatment effect, and that b is constant throughout the experiment, i.e. the relationship between x and y is the same for all treatments. It can be seen that the residual in the ordinary model $y_{ij} = M + a_i + \varepsilon_{ij}$ is reduced because part of it is accounted for by the $b(x_{ij} - \bar{x})$ and consequently the residual mean square should be smaller.

Table 11.1 gives an example of an experiment made to determine the effect of three diets on the growth of pigs during the late stages of fattening. The pigs had been marked at birth, so that one from each litter could be assigned to each treatment to form a randomized-block experiment with litters as blocks, and then run together until the experiment started. They were weighed to obtain the "initial" weight, and then fed individually the treatment diets for a standard period of time. The diets differed in quantity of protein, a_0 being the quantity recommended by authoritative advisers, a_1 $2\frac{1}{2}$ per cent less and a_2 $2\frac{1}{2}$ per cent more than the recommendation, and the null hypotheses were that a_1 does not differ from a_0, and that a_2 does not differ from a_0.

ANALYSIS OF COVARIANCE

Table 11.1—An experiment to compare three diets for bacon pigs

Live weights (lb)

Diets		Litters									Total (D)
		1	2	3	4	5	6	7	8	9	
a_0	Initial wt. (x)	161	168	184	155	162	148	151	155	161	1445
	Final wt. (y)	216	215	237	208	220	205	205	209	213	1928
a_1	Initial wt. (x)	176	159	161	169	159	162	149	156	162	1453
	Final wt. (y)	217	196	205	208	209	218	194	194	201	1842
a_2	Initial wt. (x)	162	159	174	168	154	151	155	172	149	1444
	Final wt. (y)	209	218	229	215	210	204	215	223	206	1929
Total (L)	Initial wt. (x)	499	486	519	492	475	461	455	483	472	4342
	Final wt. (y)	642	629	671	631	639	627	614	626	620	5699

Analysis of covariance

Source of variation	D.F.	S.S.$_x$	S.S.$_y$	S.P.$_{xy}$	$\frac{\text{S.P.}_{xy}^2}{\text{S.S.}_x}$	S.S.$_y - \frac{\text{S.P.}_{xy}^2}{\text{S.S.}_x}$	D.F.	M.S.	F
Litters	8	1031	732	+747					
Diets	2	5	554	−55		661	2	330	15·79
Error	16	1024	1266	+988	953	313	15	20·9	
Total	26	2060	2552	+1680					
Error + Diets	18	1029	1820	+933	846	974	17		

To analyse these data we must first work out an analysis of variance for the two variables separately in the usual way. The sum of squares for each item is shown in Table 11.1 in the columns headed S.S.$_x$ and S.S.$_y$. We also need the sum of products. In the simple case shown earlier (p. 121) it was seen that total sum of products could be derived as the sum of the products of the actual values minus the product of the two totals divided by the number of observations, it being pointed out that this procedure is exactly similar to that for getting sums of squares. Sums of products for individual components are obtained by following the same rules. Thus for litters,

$$\text{S.S.}_{\cdot x} = \frac{\sum L_x^2}{3} - \frac{(\sum x)^2}{27}$$

so

$$\text{S.P.}_{\cdot xy} = \frac{\sum L_x L_y}{3} - \frac{(\sum x)(\sum y)}{27}$$

$$= \frac{499 \times 642 + 486 \times 629 + \ldots + 472 \times 620}{3} - \frac{4342 \times 5699}{27}$$

$$= \frac{2\,751\,693}{3} - 916\,484$$

Likewise for diets,

$$\text{S.P.}_{\cdot xy} = \frac{\sum D_x D_y}{9} - \frac{(\sum x)(\sum y)}{27}$$

$$= \frac{1445 \times 1928 + 1453 \times 1842 + 1444 \times 1929}{9} - 916\,484$$

It is worth remembering that sums of products can be negative as well as positive, so they should always be given their appropriate signs when written in the table. The error sum of products is obtained by difference between the total and the sum of the other components, in the same way as are error sums of squares. At this stage it is worth testing if there is a linear relationship between the two variables, because, if there is not, then the b in the model is zero, and the error sum of squares for y will form the only reasonable basis for a test of the effects of diets. To do this we continue along the row labelled error and calculate the sum of squares of y which is due to the regression, i.e.

$$\frac{(\text{S.P.}_{\cdot xy})^2}{\text{S.S.}_{\cdot x}} = \frac{(+988)^2}{1024} = 953$$

Then in the next column we can write the sum of squares of the deviations from the regression, which is simply the whole of the sum of squares of y due to the error variation minus that due to the regression, i.e.

$$\text{S.S.}_{\cdot y} - \frac{(\text{S.P.}_{\cdot xy})^2}{\text{S.S.}_{\cdot x}} = 1266 - 953 = 313$$

This will provide the error for testing if b differs from zero, but we must decide how many D.F. it has. This is easily seen from the expression used to derive it. S.S.$_{\cdot y}$ had 16 D.F., and we have removed from S.S.$_{\cdot y}$ the sum of squares of the regression $(\text{S.P.}_{\cdot xy})^2/\text{S.S.}_{\cdot x}$ which has 1 D.F., so the residual sum of squares has $16 - 1 = 15$ D.F. and thus its mean square is $\frac{313}{15} = 20\cdot 9$. As shown on p. 122, $b = \text{S.P.}_{\cdot xy}/\text{S.S.}_{\cdot x}$ so in this case is

$$\frac{+988}{1024} = +0\cdot 96$$

and its S.E. is

$$\sqrt{\frac{\text{residual M.S.}}{\text{S.S.}_{\cdot x}}} = \sqrt{\frac{20\cdot 9}{1024}} = 0\cdot 143 \quad \text{so} \quad t = \frac{0\cdot 96}{0\cdot 143}$$

giving a value greater than 6 which is highly significant. We can conclude that b is not zero, so our model will not degrade to the original randomized-blocks model which took no account of the initial weight of the pigs.

If an F-test of the treatment effects is required, we must proceed to find the adjusted mean square for treatments. In this particular case it is not required, but we will work it out for the sake of completeness. To make the test we need to see if the variation of the treatment means from a common regression line is greater than the variation of the individual values around the line after the effect of treatment has been removed. So we need values for the S.S. of residual variation when treatment effects are included, and for the S.S. of residual effects when treatment effects are excluded, and the difference between them will be the S.S. of the treatment effects after the regression has been allowed for. Thus we produce the last row of the analysis-of-covariance table in Table 11.1 to give S.S. and S.P. for error + diets by simply adding the diet value to the error value in the first D.F. column and in the S.S.$_x$, S.S.$_y$ and S.P.$_{xy}$ columns. We then calculate (S.P.$_{xy}$)^2/S.S.$_x$ for this row, i.e.

$$\frac{(+933)^2}{1029} = 846$$

and subtract this value from the S.S.$_y$ of this row to give

$$\text{S.S.}_y - \frac{(\text{S.P.}_{xy})^2}{\text{S.S.}_x} = 1820 - 846 = 974$$

Now the residual S.S. for diets is this value minus the residual S.S. for error which we worked out previously, i.e. $974 - 313 = 661$. The degrees of freedom for this sum of squares can be determined similarly. When adding the original S.S.$_y$ of error to diet, we produced a sum of squares with 18 D.F.; from this we have subtracted the sum of squares due to a regression (S.P.$_{xy}$)^2/S.S.$_x$ so the residue has 17 D.F. The residual for the error alone has 15 D.F., as we worked out before, so that for diets has $17 - 15 = 2$ as before. This perhaps emphasizes that the regression concerned is independent of the treatment effects. The residual mean squares for treatments and error are then calculated in the usual way by dividing the sums of squares by the degrees of freedom, $\frac{661}{2} = 330$ for diets and $\frac{313}{15} = 20.9$ for error, and

$$F = \frac{\text{M.S. for diets}}{\text{M.S. for error}} = \frac{330}{20.9} = 15.79$$

in the usual way. This value of F with 2 and 15 D.F. is very highly significant, so we could conclude that, after allowing for the variation in initial weight, there is likely to be some true effect of diet. If the sum of squares for treatments needs to be broken down to provide F-tests for particular components, this procedure is applied to each component separately, e.g. if we divided our diets into orthogonal contrasts

$$\left(a_0 - \frac{a_1+a_2}{2}\right) \quad \text{and} \quad (a_1-a_2)$$

we should obtain S.S.$_x$, S.S.$_y$ and S.P.$_{xy}$ for each component separately, and add each to the error S.S. or S.P. in turn, and calculate the residual S.S. for each in the same way as we did for diets as a whole. It may be noted that if we had not made this model to include the effect of initial weight, our analysis based on the sums of squares of final weights alone S.S.$_y$ would have given us an M.S. of $\frac{554}{2} = 277$ for diets and $\frac{1266}{16} = 79 \cdot 1$ for error with

$$F = \frac{227}{79 \cdot 1} = 2 \cdot 87$$

which is much less than the 3·63 shown in the tables for $P \leqslant 0 \cdot 05$. Thus from the point of view of detecting true differences, the analysis of covariance has given us a much more precise experiment, the error mean square being only about a quarter of its value if the adjustments had not been made.

t-tests are slightly more complicated than usual but are easily worked out. First we adjust the treatment means or totals for the effects of initial live-weight. The adjustment follows from the model:

$$\bar{a}_i \text{ (adjusted)} = \bar{a}_i - b(\bar{x}_i - \bar{x})$$

so

$$\bar{a}_0 \text{ (adjusted)} = \tfrac{1}{9}\{1928 - 0 \cdot 96(1445 - \tfrac{4342}{3})\} = 214 \cdot 5$$
$$\bar{a}_1 \text{ (adjusted)} = \tfrac{1}{9}\{1842 - 0 \cdot 96(1453 - \tfrac{4342}{3})\} = 204 \cdot 1$$
$$\bar{a}_2 \text{ (adjusted)} = \tfrac{1}{9}\{1929 - 0 \cdot 96(1444 - \tfrac{4342}{3})\} = 214 \cdot 7$$

To test the first null hypothesis, that reducing the protein content of the diet does not affect the final weight, we need

$$t = \frac{\bar{a}_1 \text{ (adjusted)} - \bar{a}_0 \text{ (adjusted)}}{\text{S.E. of } \{\bar{a}_1 \text{ (adjusted)} - \bar{a}_0 \text{ (adjusted)}\}}$$

From the equations used to obtain the adjusted means,

$$\bar{a}_1 \text{ (adjusted)} - \bar{a}_0 \text{ (adjusted)} = \bar{a}_1 - b(\bar{x}_1 - \bar{x}) - \bar{a}_0 + b(\bar{x}_0 - \bar{x})$$
$$= \bar{a}_1 - \bar{a}_0 - b(\bar{x}_1 - \bar{x}_0)$$

(i.e. $\bar{x}$ cancels out) so following the usual rules the S.E. will be

$$\sqrt{[\text{S.E.}_{\bar{a}_1}^2 + \text{S.E.}_{\bar{a}_0}^2 + (\bar{x}_1 - \bar{x}_0)^2 \text{S.E.}_b^2]}$$
$$= \sqrt{\left[\text{Residual M.S. of error}\left\{\frac{1}{r} + \frac{1}{r} + \frac{(\bar{x}_1 - \bar{x}_0)^2}{\text{S.S.}_{\cdot x}}\right\}\right]}$$
$$= \sqrt{\left[20\cdot 9\left\{\frac{2}{9} + \frac{(1453-1445)^2}{9 \times 9 \times 1024}\right\}\right]} = 2\cdot 16$$

so

$$t = \frac{204\cdot 1 - 214\cdot 5}{2\cdot 16} = 4\cdot 81$$

which when compared with t in the tables with 15 D.F. is greater than that for the 0·001 level of probability. We would conclude that reducing the protein by this amount really does decrease the final weight of pigs.

The second null hypothesis (that increasing the protein does not affect final weight) would be tested in the same way by comparing $\bar{a}_2 - \bar{a}_0$ with its standard error which would be

$$\sqrt{\left[\text{M.S.}_E\left\{\frac{2}{r} + \frac{(\bar{x}_2 - \bar{x}_0)^2}{\text{S.S.}_{\cdot x}}\right\}\right]} = \sqrt{\left[20\cdot 9\left\{\frac{2}{9} + \frac{(1444-1445)^2}{9 \times 9 \times 1024}\right\}\right]} = 2\cdot 16$$

Since the difference between the two means is only 0·2, it is obvious that there is no justification for disputing the null hypothesis.

It will be seen that the S.E. of the difference between two means has to be worked out separately for each individual comparison when an analysis of covariance is used. Similarly, there is no exact common S.E. for the treatment means when they are displayed in a table, since an adjusted mean $= \bar{a}_i - b(\bar{x}_i - \bar{x})$ and its S.E. will therefore be

$$\sqrt{\left[\text{M.S.}_E\left\{\frac{1}{r} - \frac{(\bar{x}_i - \bar{x})^2}{\text{S.S.}_{\cdot x}}\right\}\right]}$$

giving a different value for each of the $\bar{a}_i$ since the $\bar{x}_i$s are different. However, Finney (1946) suggested that an approximate S.E. applicable to all means may be calculated by taking

$$\text{M.S.}_E\left(1 + \frac{\text{Treatment M.S. for } x}{\text{S.S.}_x}\right)$$

as the *effective* error mean square. The last term represents the average contribution from the xs, and the whole expression gives exact values if there are only two treatments.

The S.E. of the difference between their means by this method would be

$$\sqrt{\left[\frac{2}{r}\text{M.S.}_E\left(1+\frac{\text{Treatment M.S. for }x}{\text{S.S.}_x}\right)\right]}$$

but for two treatments,

$$\text{Treatment M.S.} = \frac{r(\bar{x}_1-\bar{x}_2)^2}{2}$$

so S.E. of the difference between the two means

$$=\sqrt{\left[\frac{2}{r}\text{M.S.}_E\left\{1+\frac{r(\bar{x}_1-\bar{x}_2)^2}{2\text{S.S.}_x}\right\}\right]}=\sqrt{\left[\text{M.S.}_E\left\{\frac{2}{r}+\frac{(\bar{x}_1-\bar{x}_2)^2}{\text{S.S.}_x}\right\}\right]}$$

Thus in the present example an approximate S.E. for the means

$$=\sqrt{\left[\frac{20\cdot 9}{9}\left(1+\frac{2\cdot 5}{1024}\right)\right]}=1\cdot 53$$

and the approximate S.E. for the two comparisons would be this value multiplied by $\sqrt{2} = 2\cdot 16$. In fact in this experiment the three S.E.s calculated are the same to two places of decimals because the $\bar{x}_i$ are all very similar.

When analysis of covariance is used to reduce the error term and give a more precise experiment, it is obvious that the xs must not themselves be affected by the treatments since, if they were then part of the treatment, effect would be taken out of the calculation as well. Thus in biology such a device can be safely used only when the x-measure can be taken before the experiment begins, i.e. before the treatments are applied. Such occasions occur frequently when growth of organisms is involved, and the size of the organisms is somewhat variable at the stage of growth at which the experiment starts and when, as is usual, growth varies as size. However, there are pitfalls to guard against.

(1) In interpretation we must be careful in extrapolating the confidence limits. We shall have limits for final weight, say, on the basis of all organisms starting at the same weight; but in real life they do not all start at the same weight, so we must not expect such small ranges in final weight of the population.

(2) It must be remembered that only the linear effect is being removed, and in many biological situations the final measure may well be affected by the initial measure in some other way, e.g. there may be a curvilinear or asymptotic relationship. Then the linear covariance will do little towards controlling the error, and further terms representing the effect of x

on y must be put into the model and, of course, the analysis becomes more complex.

(3) There is the question of whether we should use covariance routinely and not bother to remove variation by other means, such as using randomized-block designs. This depends on the relative precision obtained by the various methods. For example, in experiments with pigs, using litters as blocks takes some account of genetic variation, variation in birth weight, variation in age and variation in early nutrition and, furthermore, keeping pigs in litters leads to less fighting, tail-biting, etc., than mixing litters later in life. Thus, using a randomized-block design may often give a far more precise experiment than using analysis of covariance on only one or two of those variables. On the other hand, some biologists regularly make blocks of animals based on initial weight; if there are four treatments, say, they would take the four heaviest as Block I, the four next heaviest as Block II, and so on. In so doing, the variation between blocks may well be little different from the variation between animals in a block. In such a case analysis of covariance is often better, because it takes account of all the variation, that within blocks as well as that between blocks. In some cases we need to consider the amount of work involved. In a field experiment, for instance, it may be possible to obtain estimates of the plot variation by growing and harvesting a uniform crop, without treatments, in one year, and then apply the treatments to a second crop, the next year, using the first year's data as the xs in an analysis of covariance. Soil fertility as measured by the same crop species tends to stay fairly stable from one year to the next, so b in our analysis-of-variance model is often significant, and analysis of covariance increases the precision. However, to get this extra precision we would almost need to double the work, since we have to grow, harvest and weigh the produce from individual plots in each of two years; often equal precision could be obtained by increasing slightly the replication (not doubling it) and using an ordinary analysis of variance. With plants, analysis of covariance is likely to show the greatest benefit when the x-measure is made on the same plant as the y-measure, e.g. perennials like trees can sometimes be "calibrated" by certain measures before treatments are applied, and these measures used as the x in analysis of covariance for several years thereafter.

Another sort of variable is sometimes used in analysis of covariance for the purpose of reducing error. If, for example, the number of treatments is such that we could produce a balanced design as regards treatments and litters in an animal experiment, but could not produce balance for sex as well, to give a three-factor design, we might give the sexes values of 0 for males and 1 for females, and use these as xs in an analysis of covariance. Interpretation needs care since, suppose in the whole experiment there were 18 males and 22 females, $\bar{x}$ would be

$$\frac{18 \times 0 + 22 \times 1}{40} = 0\cdot 55$$

so, after adjustment in the usual way of $\bar{a}_i - b(\bar{x}_i - \bar{x})$, we should have results appropriate to groups of animals in which 55 per cent were female, not about these animals in general which might well have something nearer half and half of males and females. Of course, the adjustment should be made by replacing $\bar{x}$ by 0·5 or whatever is the known proportion. Similarly, we may sometimes use analysis of covariance when an accident has occurred during an experiment and certain items have been damaged; although they are providing some data, we cannot be sure that they are truly representative of the treatments they are meant to estimate. Again, making $x = 0$ when there was no damage and $x = 1$ when there was, carrying out an analysis of covariance will show first if the damage has affected the final measure (i.e. if b is significant) and, if it has, the final measures can be adjusted. In this case we must be very careful to ensure that treatments themselves have not effected the damage, or much of the information given by the experiment will have been discarded.

Analysis of covariance is sometimes used for another purpose, namely, to sort out the various ways in which the treatment is effective, but when doing this great care is required in interpretation to avoid unjustified extrapolations. Suppose we have an experiment testing the effects of certain seed-treatments on the yield of sugar beet. The treatments could affect yield in two ways: by affecting the number of plants per unit area, and by affecting the growth of these plants; but in any case the number of plants per unit area might affect the yield. At harvest we could count the number of plants on each plot (x) and obtain the yield of each plot (y). Now in an analysis of covariance we might first find that the treatment M.S. for x was significant, that for y was significant, but that the M.S. for adjusted treatments was not. We can then conclude that the effect of the treatments on yield is simply a reflection of their effect on number of plants. When we have adjusted the yield data for the differing numbers of plants, we have removed the whole of the treatment effect. It would, however, be silly to conclude that treatments have no effect on yield provided a particular plant density is maintained, because the treatments themselves dictate that a particular plant density is not maintained.

Alternatively the F-test for the M.S. for adjusted treatments might still be significant. In this case we know that all the effects of treatments cannot be attributed to the linear effect of number of plants alone, but we cannot be much more definite than that. The remaining effects could be due to non-linear effects of number of plants or to some interaction between treatments and number of plants, so in such a situation the adjusted treatment means often have very little meaning. Before using analysis of covariance for this purpose, it is as well to satisfy ourselves that a linear effect between the two measures is realistic. Also it is of great importance that the slope of the regression should be the same for all treatments; otherwise wrong answers can be obtained.

FURTHER READING

R. L. Anderson, 1946, "Missing-plot Techniques," *Biometrics Bulletin* **2**, 41–47.
J. G. Bald, 1943, "Estimation of the Leaf Area of Potato Plants for Pathological Studies," *Phytopathology* **33**, 922–32.
W. G. Cochran and Gertrude M. Cox, 1957, *Experimental Designs*, New York, London and Sydney, Wiley, 2 ed.
W. J. Dixon and F. J. Massey, 1957, *Introduction to Statistical Analysis*, New York, Toronto and London, McGraw-Hill.
D. B. Duncan, 1955, "Multiple Range and Multiple F-Tests," *Biometrics* **11**, 1.42.
C. W. Dunnett, 1955, "A Multiple Comparison Procedure for Comparing Several Treatments with a Control," *American Statistical Association Journal* **50**, 1096–121.
D. J. Finney, 1946, "Standard Errors of Yields Adjusted for Regression on an Independent Measurement," *Biometrics Bulletin* **2**, 53–55.
R. A. Fisher, 1925, *Statistical Methods for Research Workers*, Edinburgh, Oliver and Boyd.
Sir R. A. Fisher and F. Yates, 1963, *Statistical Tables*, Edinburgh, Oliver and Boyd, 6 ed.
D. J. Greenwood, T. J. Cleaver and K. B. Niendorf, 1974, "Effects of Weather Conditions on Response of Lettuce to Applied Fertilizer," *Journal of Agricultural Science, Cambridge* **82**, 217–32.
K. Mather, 1964, *Statistical Analysis in Biology*, London, Methuen, 5 ed.
M. R. Sampford, 1962, *An Introduction to Sampling Theory*, Edinburgh, Oliver and Boyd.
S. Siegel, 1956, *Nonparametric Statistics for the Behavioral Sciences*, New York, Toronto and London, McGraw-Hill.
G. W. Snedecor, 1934, *Analysis of Variance and Covariance*, Ames, Collegiate Press.
R. R. Sokal and F. Rohlf, 1969, *Biometry*, San Francisco, Freeman.
"Student," 1908, "The Probable Error of a Mean," *Biometrika*, vi, 1.25.
F. Yates, 1933, "The Analysis of Replicated Experiments when the Field Results are Incomplete," *Empire Journal of Experimental Agriculture* **1**, 129–42.
F. Yates, 1934, "Contingency Tables Involving Small Numbers and the χ^2 Tests," *Supplement to the Journal of the Royal Statistical Society* **1**, 217.
F. Yates, 1937, "The Design and Analysis of Factorial Experiments," *Imperial Bureau of Soil Science. Technical Communication No. 35*.
F. Yates, 1960, *Sampling Methods for Censuses and Surveys*, London, Griffin, 3 ed.

APPENDIX

TABLE I. Random digits

15	06	66	52	85	08	91	08	42	19	41	93
49	99	63	50	82	88	68	86	97	15	98	25
98	39	41	25	75	11	36	63	72	15	33	58
26	96	87	01	89	87	63	07	05	18	13	85
58	32	98	94	80	38	45	63	38	30	51	29
82	12	95	14	10	92	26	41	13	70	69	00
31	23	80	99	89	00	58	09	25	48	36	45
23	97	31	11	90	13	42	00	13	98	94	54
00	46	52	81	15	87	19	56	31	03	11	64
24	97	18	70	53	36	60	79	87	44	61	93
59	74	27	52	03	45	69	22	10	84	52	24
09	85	71	01	15	41	62	62	14	93	91	83
69	97	47	96	48	54	31	62	11	57	23	82
78	93	01	88	85	25	38	14	72	96	70	98
48	22	92	21	40	40	98	66	11	57	23	78
10	13	17	51	43	01	85	02	06	11	55	89
91	89	70	58	22	69	19	92	88	21	80	04
12	61	01	06	93	24	56	26	83	25	79	31
94	81	80	36	79	71	57	77	17	38	79	84
91	86	61	44	18	75	96	03	87	90	40	64
76	50	17	57	25	79	59	69	65	29	50	85
50	52	40	43	33	29	95	93	37	08	20	83
53	66	15	03	46	83	86	99	57	55	41	13
36	02	45	95	35	75	67	69	70	18	98	72
01	44	26	54	50	09	80	29	85	45	08	84
28	83	17	25	77	91	29	25	51	48	08	23
79	50	31	09	52	11	11	39	03	97	91	09
19	42	82	25	78	69	38	79	10	16	01	38
09	46	36	76	02	11	03	61	29	11	57	74
75	56	01	00	99	89	40	71	84	05	30	68

TABLE II. Values of t and F. In each cell the first value is the 5 per cent ($P = 0.05$), the second the 1 per cent ($P = 0.01$), and the third the 0.1 per cent ($P = 0.001$) level of probability

D.F. for Error M.S. (n_2)	t	F D.F. for Treatments M.S. (n_1)						
		1	2	3	4	5	6	8
5	2.57 4.03 6.86	6.61 16.26 47.04	5.79 13.27 36.61	5.41 12.06 33.20	5.19 11.39 31.09	5.05 10.97 29.75	4.95 10.67 28.84	4.82 10.27 27.64
6	2.45 3.71 5.96	5.99 13.74 35.51	5.14 10.92 27.00	4.76 9.78 23.70	4.53 9.15 21.90	4.39 8.75 20.81	4.28 8.47 20.03	4.15 8.10 19.03
7	2.36 3.50 5.41	5.59 12.25 29.22	4.74 9.55 21.69	4.35 8.45 18.77	4.12 7.85 17.19	3.97 7.46 16.21	3.87 7.19 15.52	3.73 6.84 14.63
8	2.31 3.36 5.04	5.32 11.26 25.42	4.46 8.65 18.49	4.07 7.59 15.83	3.84 7.01 14.39	3.69 6.63 13.49	3.58 6.37 12.86	3.44 6.03 12.05
9	2.26 3.25 4.78	5.12 10.56 22.86	4.26 8.02 16.39	3.86 6.99 13.90	3.63 6.42 12.50	3.48 6.06 11.71	3.37 5.80 11.13	3.23 5.47 10.37
10	2.23 3.17 4.59	4.96 10.04 21.04	4.10 7.56 14.91	3.71 6.55 12.55	3.48 5.99 11.28	3.33 5.64 10.48	3.22 5.39 9.92	3.07 5.06 9.20
11	2.20 3.11 4.44	4.84 9.65 19.68	3.98 7.20 13.81	3.59 6.22 11.56	3.36 5.67 10.35	3.20 5.32 9.58	3.09 5.07 9.05	2.95 4.74 8.35
12	2.18 3.05 4.32	4.75 9.33 18.64	3.88 6.93 12.97	3.49 5.95 10.81	3.26 5.41 9.63	3.11 5.06 8.89	3.00 4.82 8.38	2.85 4.50 7.71

APPENDIX

D.F. for Error M.S. (n_2)	t	F D.F. for Treatments M.S. (n_1)						
		1	2	3	4	5	6	8
13	2·16	4·67	3·80	3·41	3·18	3·02	2·92	2·77
	3·01	9·07	6·70	5·74	5·20	4·86	4·62	4·30
	4·22	17·81	12·31	10·21	9·07	8·35	7·86	7·21
14	2·14	4·60	3·74	3·34	3·11	2·96	2·85	2·70
	2·98	8·86	6·51	5·56	5·03	4·69	4·46	4·14
	4·14	17·14	11·78	9·73	8·62	7·92	7·43	6·80
15	2·13	4·54	3·68	3·29	3·06	2·90	2·79	2·64
	2·95	8·68	6·36	5·42	4·89	4·56	4·32	4·00
	4·07	16·58	11·34	9·34	8·25	7·57	7·09	6·47
16	2·12	4·49	3·63	3·24	3·01	2·85	2·74	2·59
	2·92	8·53	6·23	5·29	4·77	4·44	4·20	3·89
	4·02	16·12	10·97	9·00	7·94	7·27	6·80	6·20
18	2·10	4·41	3·55	3·16	2·93	2·77	2·66	2·51
	2·88	8·28	6·01	5·09	4·58	4·25	4·01	3·71
	3·92	15·38	10·39	8·49	7·46	6·81	6·35	5·76
20	2·09	4·35	3·49	3·10	2·87	2·71	2·60	2·45
	2·85	8·10	5·85	4·94	4·43	4·10	3·87	3·56
	3·85	14·82	9·95	8·10	7·10	6·46	6·02	5·44
22	2·07	4·30	3·44	3·05	2·82	2·66	2·55	2·40
	2·82	7·94	5·72	4·82	4·31	3·99	3·75	3·45
	3·79	14·38	9·61	7·80	6·81	6·19	5·76	5·19
24	2·06	4·26	3·40	3·01	2·78	2·62	2·51	2·36
	2·80	7·82	5·61	4·72	4·22	3·90	3·67	3·36
	3·74	14·03	9·34	7·55	6·59	5·98	5·55	4·99
30	2·04	4·17	3·32	2·92	2·69	2·53	2·42	2·27
	2·75	7·56	5·39	4·51	4·02	3·70	3·47	3·17
	3·65	13·29	8·77	7·05	6·12	5·53	5·12	4·58
60	2·00	4·00	3·15	2·76	2·52	2·37	2·25	2·10
	2·66	7·08	4·98	4·13	3·65	3·34	3·12	2·82
	3·46	11·97	7·77	6·17	5·31	4·76	4·37	3·87
∞	1·96	3·84	3·00	2·60	2·37	2·21	2·09	1·94
	2·58	6·64	4·62	3·78	3·32	3·02	2·80	2·51
	3·29	10·83	6·91	5·42	4·62	4·10	3·74	3·27

TABLE III. Values of χ^2

D.F.	\multicolumn{8}{c}{Probability level (P)}							
	0·99	0·975	0·95	0·90	0·10	0·05	0·025	0·01
1	0·00	0·00	0·00	0·02	2·71	3·84	5·02	6·64
2	0·02	0·05	0·10	0·21	4·60	5·99	7·38	9·21
3	0·12	0·22	0·35	0·58	6·25	7·82	9·35	11·34
4	0·30	0·48	0·71	1·06	7·78	9·49	11·14	13·28
5	0·55	0·83	1·14	1·61	9·24	11·07	12·83	15·09
6	0·87	1·24	1·64	2·20	10·64	12·59	14·45	16·81
8	1·65	2·18	2·73	3·49	13·36	15·51	17·54	20·09
10	2·56	3·25	3·94	4·86	15·99	18·31	20·48	23·21

TABLE IV. Values for Duncan's Multiple Range Test. The upper value in each cell is for testing at the 5 per cent ($P = 0·05$) and the lower value for testing at the 1 per cent ($P = 0·01$) level of probability

D.F. for Error M.S. (n_2)	\multicolumn{6}{c}{Distance apart in the range (p)}					
	2	3	4	5	6	8	10
10	3·15	3·29	3·37	3·43	3·46	3·47	3·47
	4·48	4·73	4·88	4·96	5·06	5·20	5·28
12	3·08	3·23	3·33	3·36	3·40	3·44	3·46
	4·32	4·55	4·68	4·76	4·84	4·96	5·07
14	3·03	3·18	3·27	3·33	3·37	3·41	3·44
	4·21	4·42	4·55	4·63	4·70	4·83	4·91
16	3·00	3·15	3·23	3·30	3·34	3·39	3·43
	4·13	4·34	4·45	4·54	4·60	4·72	4·79
18	2·97	3·12	3·21	3·27	3·32	3·37	3·41
	4·07	4·27	4·38	4·46	4·53	4·64	4·71
20	2·95	3·10	3·18	3·25	3·30	3·36	3·40
	4·02	4·22	4·33	4·40	4·47	4·58	4·65
24	2·92	3·07	3·15	3·22	3·28	3·34	3·38
	3·96	4·14	4·24	4·33	4·39	4·49	4·57

TABLE V. Values of t for joint confidence at 95 per cent (upper value) and 99 per cent (lower value) when comparing several treatments with a control. After Dunnett (1955)

(a) Two-sided comparisons

D.F. of Error M.S. (n_2)	\multicolumn{9}{c}{Number of treatments excluding the control (p)}								
	1	2	3	4	5	6	7	8	9
10	2·23	2·57	2·81	2·97	3·11	3·21	3·31	3·39	3·46
	3·17	3·53	3·78	3·95	4·10	4·21	4·31	4·40	4·47
11	2·20	2·53	2·76	2·92	3·05	3·15	3·24	3·31	3·38
	3·11	3·45	3·68	3·85	3·98	4·09	4·18	4·26	4·33
12	2·18	2·50	2·72	2·88	3·00	3·10	3·18	3·25	3·32
	3·05	3·39	3·61	3·76	3·89	3·99	4·08	4·15	4·22
13	2·16	2·48	2·69	2·84	2·96	3·06	3·14	3·21	3·27
	3·01	3·33	3·54	3·69	3·81	3·91	3·99	4·06	4·13
14	2·14	2·46	2·67	2·81	2·93	3·02	3·10	3·17	3·23
	2·98	3·29	3·49	3·64	3·75	3·84	3·92	3·99	4·05
15	2·13	2·44	2·64	2·79	2·90	2·99	3·07	3·13	3·19
	2·95	3·25	3·45	3·59	3·70	3·79	3·86	3·93	3·99
16	2·12	2·42	2·63	2·77	2·88	2·96	3·04	3·10	3·16
	2·92	3·22	3·41	3·55	3·65	3·74	3·82	3·88	3·93
17	2·11	2·41	2·61	2·75	2·85	2·94	3·01	3·08	3·13
	2·90	3·19	3·38	3·51	3·62	3·70	3·77	3·85	3·89
18	2·10	2·40	2·59	2·73	2·84	2·92	2·99	3·05	3·11
	2·88	3·17	3·35	3·48	3·58	3·67	3·74	3·80	3·85
19	2·09	2·39	2·58	2·72	2·82	2·90	2·97	3·04	3·09
	2·86	3·15	3·33	3·46	3·55	3·64	3·70	3·76	3·81
20	2·09	2·38	2·57	2·70	2·81	2·89	2·96	3·02	3·07
	2·85	3·13	3·31	3·43	3·53	3·61	3·67	3·73	3·78
24	2·06	2·35	2·53	2·66	2·76	2·84	2·91	2·96	3·01
	2·80	3·07	3·24	3·36	3·45	3·52	3·58	3·64	3·69

TABLE V. (continued)
(b) One-sided comparisons

D.F. of Error M.S. (n_2)	\multicolumn{9}{c}{Number of treatments excluding the control (p)}								
	1	2	3	4	5	6	7	8	9
10	1·81 2·76	2·15 3·11	2·34 3·31	2·47 3·45	2·56 3·56	2·64 3·64	2·70 3·71	2·76 3·78	2·81 3·83
11	1·80 2·72	2·13 3·06	2·31 3·25	2·44 3·38	2·53 3·48	2·60 3·56	2·67 3·63	2·72 3·69	2·77 3·74
12	1·78 2·68	2·11 3·01	2·29 3·19	2·41 3·32	2·50 3·42	2·58 3·50	2·64 3·56	2·69 3·62	2·74 3·67
13	1·77 2·65	2·09 2·97	2·27 3·15	2·39 3·27	2·48 3·37	2·55 3·44	2·61 3·51	2·66 3·56	2·71 3·61
14	1·76 2·62	2·08 2·94	2·25 3·11	2·37 3·23	2·46 3·32	2·53 3·40	2·59 3·46	2·64 3·51	2·69 3·56
15	1·75 2·60	2·07 2·91	2·24 3·08	2·36 3·20	2·44 3·29	2·51 3·36	2·57 3·42	2·62 3·47	2·67 3·52
16	1·75 2·58	2·06 2·88	2·23 3·05	2·34 3·17	2·43 3·26	2·50 3·33	2·56 3·39	2·61 3·44	2·65 3·48
17	1·74 2·57	2·05 2·86	2·22 3·03	2·33 3·14	2·42 3·23	2·49 3·30	2·54 3·36	2·59 3·41	2·64 3·45
18	1·73 2·55	2·04 2·84	2·21 3·01	2·32 3·12	2·41 3·21	2·48 3·27	2·53 3·33	2·58 3·38	2·62 3·42
19	1·73 2·54	2·03 2·83	2·20 2·99	2·31 3·10	2·40 3·18	2·47 3·25	2·52 3·31	2·57 3·36	2·61 3·40
20	1·72 2·53	2·03 2·81	2·19 2·97	2·30 3·08	2·39 3·17	2·46 3·23	2·51 3·29	2·56 3·34	2·60 3·38
24	1·71 2·49	2·01 2·77	2·17 2·92	2·28 3·03	2·36 3·11	2·43 3·17	2·48 3·22	2·53 3·27	2·57 3·31

Index

analysis of covariance   214–223
analysis of variance   56–58, 132, 137, 145, 156

χ^2-test   83–86
coding data   25–26
coefficient of determination   128
coefficient of variation   33–34, 59
confidence limits   40–42, 52–53, 76–77, 124–126, 135–136
confounding in
    2^3 experiments   166–172
    2^5 experiments   173–176
    2^6 experiments   176–178
    3^2 experiments   182–184
    3^3 experiments   184–186
    $3 \times 2 \times 2$ experiments   188–193
    $3 \times 3 \times 2$ experiments   193
    4^2 experiments   187, 194
control treatment   66–71
correction factor   24
correlation   126–129
curve fitting   107–108, 111–114, 124

degrees of freedom   27–28, 31–32
design   12–16, 43–44
deviates   23
dispersion   22
Duncan's multiple range test   64–66
Dunnett's test   66–69

error variation   130–137
expected mean squares   132, 133, 137, 145

factorial design 81–100
Fisher-Behrens test 48–49
fixed-effect factor 131, 136–137
fractional replication 179–182
frequency distribution 17–22
Friedmann procedure 143–144
F-test 58–61, 132

hierarchical design 133–135

interaction 82, 95–96, 108–119, 136–137
interval-scale data 4

Kendall's rank correlation 128–129
Kruskal-Wallis procedure 61–62

Latin square 147–154, 201–207
lattice designs 195–201
least significant difference 39
location 19

main effect 82
Mann-Whitney test 49–52
mean 19
mean square 27
median 19
missing values 212–213
mode 19
models 48, 130–137, 208–212, 214

nominal data 3
non-orthogonal analysis 208–212
normal distribution 21–22, 35
null hypothesis 13

ordinal data 4
orthogonal contrasts 71–80
orthogonal polynomials 101–114

partial confounding 167–172, 188–194
probability 36

random-effect factors 131, 136–137
randomization 9–12
randomized-block design 137–141
range 22
ranking 4, 50–51, 61, 128, 141–142
regression 119–126

replication   13, 44–46, 69–71
rounding   40

samples   8–12
significance   38
Sign-test   142–143
single-factor design   46–48
single replicate experiments   173–178
skewed distribution   19–20
split-plot designs   155–165
standard deviation   29
standard error
    for split-plot designs   160–161
    in analysis of covariance   219–220
    in lattice squares   201
    of a difference   32–33
    of a linear contrast   76–77
    of a mean   29–31
    of a single observation   29
sum of products   121–122
sum of squares   23–26, 57, 73–75, 117, 130–131, 191

three-factor design   145–147
transformation of data   6–7
treatment   14
t-test   36–37, 60–61, 63, 76, 161
two-factor design   136–144

unbiased estimates   21–22, 31–32

variance   27–32, 132–137

Wilcoxon's signed-rank test   141–142